T0335978

BIOPRINTING
Principles and Applications

BIOPRINTING
Principles and Applications

Chee Kai Chua • Wai Yee Yeong
Nanyang Technological University, Singapore

 **World Scientific**

NEW JERSEY • LONDON • SINGAPORE • BEIJING • SHANGHAI • HONG KONG • TAIPEI • CHENNAI

Published by

World Scientific Publishing Co. Pte. Ltd.
5 Toh Tuck Link, Singapore 596224
USA office: 27 Warren Street, Suite 401-402, Hackensack, NJ 07601
UK office: 57 Shelton Street, Covent Garden, London WC2H 9HE

Library of Congress Cataloging-in-Publication Data
Chua, Chee Kai, author.
  Bioprinting : principles and applications / Chua Chee Kai, Yeong Wai Yee.
  p. ; cm.
  Includes bibliographical references and index.
  ISBN 978-9814612104 (hardbound : alk. paper) -- ISBN 9814612103 (hardbound : alk. paper)
  I. Yeong, Wai Yee, author. II. Title.
  [DNLM: 1. Bioprinting. QT 36]
  RB113
  616.07--dc23
                        2014039251

British Library Cataloguing-in-Publication Data
A catalogue record for this book is available from the British Library.

Printed in Singapore

Dedication

My wife, Wendy, and children, Cherie, Clement and Cavell, whose forbearance, support and motivation have made it possible for us to finish writing this book.

Chee Kai

Tee Seng. For his faith, and our bundles of joy, Bao Rong and Zi Kai.

Wai Yee

Foreword

Three-dimensional (3D) bioprinting, also known as organ printing, is a rapidly evolving multidisciplinary area of biomedical research and it can be defined as a robotic additive biofabrication of functional 3D tissue and organ constructs using biomaterials and living cells according to the predefined digital model. Publication of a textbook on 3D Bioprinting is an important historic event and milestone in the development of 3D bioprinting technology. It is a direct manifestation of the beginning of maturation of this emerging research and engineering field.

Outstanding textbooks are usually written by prestigious academics who have extensive knowledge on the field, have significant experience in teaching for students at universities and have already written and published books. In this context, Profs. Chee Kai Chua and Wai Yee Yeong's team represents: (i) a research group with impressive record of highly cited publications in top professional journals; (ii) one of the first in the world to conduct a special Masters course on "Biofabrication" including 3D bioprinting at Nanyang Technological University; and (iii) having Prof. Chee Kai Chua as already the leading author of one of the best textbooks on "Rapid Prototyping" in the first 3 editions and the latest edition "3D Printing and Additive Manufacturing" released in 2014, with all 4 editions being very popular with students, professors and engineers.

I am sure that this very well organised and written, professionally balanced and richly illustrated book on 3D bioprinting will be another textbook that will be a classic in the bioprinting field. It will be highly demanded by students, engineers, researchers and educators who have the courage and ambitions to explore the field of bioprinting. Moreover, writing and publishing such a textbook on 3D bioprinting and the associated organisation, classification and presentation of accumulated new knowledge is also an important contribution to the further development of this research field. I am also very confident that the researcher who will first successfully bioprint and implant human organ will say in his or her Nobel speech that he or she was inspired by Profs. Chee Kai Chua and Wai Yee Yeong's classic textbook on 3D bioprinting.

Vladimir Mironov

MD, PhD

Visting Professor

Division of 3D Technologies

Renato Archer Center for Information Technology

Campinas, SP, Brazil

&

Chief Scientific Officer

Laboratory of Biotechnological Research

3D Bioprinting Solutions,

Moscow, Russia

Preface

With the ever increasing demand for suitable replacements and organ transplantation, Tissue Engineering (TE), an interdisciplinary technology which emerged thirty years ago, has become a feasible solution, bringing great hope to patients who are desperate to look for tissue and organ substitutes. The classic approaches are solid scaffold-based biofabrication approaches, which utilise a temporary and biodegradable supporting structure, known as scaffold, for engineering and culturing specific tissues. These approaches, though promising, still face a number of challenges, including immunogenicity, host inflammatory responses, rate of degradation, toxicity of degradation products and fibrous tissue formation resulting from scaffold degradation. These issues may have long term effects for the biological function of the engineered tissue.

In recent years, three-dimensional (3D) bioprinting has drawn increasing attention due to its ability to create 3D structures with living biological elements. Many scientists and engineers believe that 3D bioprinting has the potential to emerge as the leading manufacturing paradigm of the 21st century. Bioprinting requires a broad range of expertise from three major disciplines, namely, biology (e.g. tissue and cell behaviours), mechanical engineering (e.g. additive manufacturing, machine design and control and CAD/CAM) and materials science (e.g. biomaterials). Working in this field requires extensive interdisciplinary knowledge. As a result, students and engineers who are from other areas often find difficult and sometimes frustrating to understand the principles of bioprinting.

This textbook, as the first textbook in 3D bioprinting, is thus written to bridge the gaps between the abovementioned three disciplines, providing not only the fundamentals, but also the insights to students and engineers. This book starts with the introduction of tissue engineering and the scaffold-based TE approaches. This is followed by four cohesive chapters elaborating the three key stages in 3D bioprinting, which are pre-processing (biomaterials and cell source), processing (the 3D bioprinting systems and processes) and post-processing (cell culture). The book also has a special chapter introducing an interesting area, namely, computer modelling and simulation for bioprinting. In particular, this book describes the concepts of tissue fusion and fluidity in great detail, which are fundamental for the modern 3D printing technology. To be used more effectively for both undergraduate and postgraduate students in Mechanical, Biomedical, Production or Manufacturing Engineering, this book provides a number of problems specifically designed based on the chapter context, which aims to highlight the key points in each chapter. For university professors and lecturers, the subject bioprinting can be easily combined and used along with other topics in mechanical, manufacturing, biomechanical and biomedical areas.

Chua C. K.
Professor

Yeong W. Y.
Assistant Professor

School of Mechanical and Aerospace Engineering
Nanyang Technological University
50 Nanyang Avenue
Singapore 639798

Acknowledgements

First, we would like to thank God for granting us His strength throughout the writing of this book. Secondly, we are especially grateful to our respective spouses, Wendy and Tee Seng, and our respective children, Cherie, Clement, Cavell Chua and Bao Rong, Zi Kai Lim for their patience, support and encouragement throughout the year it took to complete this edition.

We wish to thank the valuable support from the administration of Nanyang Technological University (NTU), especially the School of Mechanical and Aerospace Engineering (MAE) in particular the NTU Additive Manufacturing Centre (NAMC). In particular, we would like to express sincere appreciation to our research fellow, Dr Zicheng Zhu, for his invaluable input and contributions throughout the entire book. We would also like to thank Dr Jia An and our special assistant Shih Zoe Chu for their selfless help and immense effort in the coordination and timely publication of the book.

In addition, we would like to thank our colleagues and former students, Associate Professor Kah Fai Leong, Dr. May Win Naing, Dr. Florencia Edith Wiria, Dr Novella Sudarmadji, Dr Jia Yong Tan, Dr Jolene Liu, Dr Dan Liu, Ker Chin Ang, Kwang Hui Tan and Althea Chua. Credit also goes to Dr. M. Chandrasekeran, Dr Shoufeng Yang and Dr. Cheah Chi Mun.

We would also like to extend our special appreciation to Prof. Vladimir Mironov for his foreword. The acknowledgements would not be complete without the contributions of the following companies for supplying and helping us with the information about their products they develop, manufacture or represent:

1	3D Biomatrix Inc., USA
2	Bio Med Sciences, Inc., USA
3	Bio Scaffold International Ltd., Singapore
4	Cyfuse Biomedical K.K., Japan
5	Digilab Inc., USA
6	Elsevier Ltd., UK
7	EnvisionTEC GmbH, Germany
8	Fujifilm Ltd, Japan
9	GeSim mbH, Germany
10	InSphero Inc., USA
11	John Wiley & Sons, Inc., USA
12	LifeCell Corporation, USA
13	MedSkin Solutions Dr. Suwelack AG, Germany
14	Microjet Corporation, Japan
15	nScrypt, Inc., USA
16	Organovo Inc., USA
17	Osteopore Int Pte Ltd., Singapore
18	RegenHU Ltd., Switzerland
19	Royal Society of Chemistry, UK
20	Springer Science+Business Media, Germany
21	Synthecon Inc., USA
22	Tissue Regeneration Systems Inc., USA

Last but not least, we also wish to express our thanks and apologies to the many others not mentioned above for their suggestions, corrections and contributions to the success of the previous editions of the book. We would appreciate your comments and suggestions on this book.

Chua C. K.
Professor

Yeong W. Y.
Assistant Professor

About the Authors

Chee Kai CHUA is a Professor of School of Mechanical and Aerospace Engineering and Director of the NTU Additive Manufacturing Centre (NAMC) at Nanyang Technological University. Over the last 25 years, Prof Chua has established a strong research group at NTU pioneering and leading in computer-aided tissue engineering scaffold fabrication using various additive manufacturing techniques. He is internationally recognized for his significant contributions in the bio-material analysis and rapid prototyping process modelling and control for tissue engineering. His work has since extended further into additive manufacturing of metals and ceramics for defence applications.

Prof Chua has published extensively with over 200 international journal and conference papers attracting over 4300 citations and a Hirsch index of 32 in Web of Science. His book, into the 4th edition, entitled "3D Printing and Additive Manufacturing: Principles and Applications", is widely used in American, European and Asian universities and is acknowledged by international academics as one of the best textbooks in the field. He is the World No. 1 Author for the area in 'Rapid Prototyping' or '3D Printing' in Web of Science and is the most 'Highly Cited Scientist' in the world for that topic. He is the Co-Editor in Chief of the International Journal of Virtual & Physical Prototyping and serves as editorial board member of 3 other international journals. As a dedicated educator who is passionate in training the next generation, Prof Chua is widely consulted on additive manufacturing since 1990 and has conducted numerous professional development courses for mechanical engineers in Singapore and the region. In 2013, Prof Chua was awarded the "Academic Career Award" for his contributions to Additive

Manufacturing (or 3D Printing) at the 6th International Conference on Advanced Research in Virtual and Rapid Prototyping (VRAP 2013), 1 – 5 October, 2013, at Leiria, Portugal. Dr Chua can be contacted by email at mckchua@ntu.edu.sg.

Wai Yee YEONG is an assistant professor at the School of Mechanical and Aerospace Engineering, Nanyang Technological University, Singapore. She is the Deputy Director (Development) at the NTU Additive Manufacturing Centre. She has developed and taught Biofabrication, a Master course at the school. She is the associate editor of the Virtual and Physical Prototyping Journal. She has delivered invited lectures at conferences and institutes in 3D printing and bioprinting. Her research interests include bioprinting, tissue engineering and scaffold, 3D printing and development of medical devices. She has won academic prizes including best paper awards at several international conferences. Dr Yeong can be contacted by email at wyyeong@ntu.edu.sg

List of Abbreviations

2D	Two-Dimensional
3D	Three-Dimensional
3DP	Three-Dimensional Printing
ACL	Anterior Cruciate Ligament
BioLP	Biological Laser Printing
BMP-2	Bone Morphogenetic Protein-2
AM	Additive Manufacturing
B-Rep	Boundary Representation
CAD	Computer-Aided Design
CAM	Computer-Aided Manufacturing
CATE	Computer-Aided Tissue Engineering
CASTS	Computer-Aided System for Tissue Scaffolds
CFD	Computational Fluid Dynamics
CP	Cellular Particle
CPD	Cellular Particle Dynamics
CPJ	Colour Jet Printing
CSG	Constructive Solid Geometry
CT	Computed Tomography
DAH	Differential Adhesion Hypothesis
DMF	N,N-dimethyl Formamide
DmFb	Canine Dermal Fibroblasts
ECM	Extracellular Matrices
EHDJ	Electrohydrodynamic Jetting
EGFR	Epidermal Growth Factor Receptor
EtO	Ethylene Oxide
FDA	Food and Drug Administration
FDM	Fused Deposition Modelling

FEA	Finite Element Analysis
FEP	Fluorinated Ethylene Propylene
FGS	Functionally Graded Scaffold
gelMA	Gelatin Methacrylamide
GAG	Glycosaminoglycans
GUI	Graphical User Interface
HA	Hyaluronic acid
HAP	Hydroxyapatite
HBSS	Hank's Balanced Salt Solution
HepG2	Hepatocarcinoma Cell Line
hESC	Embryonic Stem Cell
HIPE	High-Internal-Phase Emulsion
HIV	Human Immunodeficiency Virus
hMSC	Human Mesenchymal Stem Cell
HPMAm	N-(2-hydroxypropyl) Methacrylamide
HSC	Haematopoietic Progenitor/Stem Cell
IGES	Initial Graphics Exchange Specification
iPSC	Induced Pluripotent Stem Cell
ISM	Implicit Surface Modelling
ITO	Indium Tin Oxide
KMC	Kinetic Monte Carlo
LCST	Lower Critical Solution Temperature
LDM	Low-temperature Deposition Manufacturing
LGDW	Laser Guidance Direct Write
LIFT	Laser Induced Forward Transfer
MCH	Monte Carlo Simulations
MCTS	Multicellular Tumour Spheroids
MDM	Multi-nozzle Deposition Manufacturing
MMC	Metropolis Monte Carlo
MVEC	Microvascular Epithelial Cells
MRI	Magnetic Resonance Imaging
MSCs	Mesenchymal Stromal/Stem Cells
nHA	nano-Hydroxyapatite
OOC	Organ On a Chip
OPF	Oligo (poly(ethylene glycol) fumarate)
PAM	Pressure-Assisted Microsyringes

PCL	Poly(caprolactone)
PDMS	Poly(dimethylsiloxane)
PE	Polyethylene
PED	Precision Extruding Deposition
PEG	Poly(ethylene glycol)
PEGDMA	Poly(ethylene glycol) Dimethacrylate
PEM	Precise Extrusion Manufacturing
PEO	Poly(ethylene oxide)
PGA	Poly(glycolic acid)
PHEMA	Poly(2-hydroxethyl methacrylate)
PIPAAm or PNIPAAm	Poly(N-isopropylacrylamide)
PLCL	Poly(lactic acid-co-(ε-caprolactone))
PLGA	Poly(lactic-co-glycolic) Acid
PLLA	Poly-L-lactic Acid
P(NIPAAm-co-HEMA)	Poly(N-isopropylacrylamide-co-hydroxylethyl methacrylate)
PP-TCP	Polypropylene-Tricalcium Phosphate
PPF	Poly(propylene fumarate)
PTFE	Polytetrafluoroethylene
PU	Polyurethane
PVA	Poly(vinyl alcohol)
RP	Rapid Prototyping
RM	Rapid Manufacturing
SAL	Sterility Assurance Limit
SCPL	Solvent Casting and Particulate Leaching
SFF	Solid Freeform Fabrication
SLA	Stereolithography Apparatus
SLS	Selective Laser Sintering
STL	Stereolithography
TE	Tissue Engineering
TEMED	Tetramethyl Ethylenediamine
T_g	Glass Transition Temperature
T_m	Melting Temperature
UV	Ultraviolet
VSMC	Vascular Smooth Muscle Cells

Contents

Chapter 1

Introduction to Tissue Engineering

Tissue engineering (TE), an interdisciplinary technology which emerged thirty years ago, has increasingly drawn significant attention from numerous scientists, engineers, technologists and physicians around the world due to its ability to construct biological substitutes to repair and replace diseased and damaged tissues. Tissue engineering technology has already been used in clinical applications, developing effective therapies, particularly for skin, blood vessels and livers. An important branch of tissue engineering has been the incorporation of new techniques such as microfabrication and additive manufacturing that will enable future breakthroughs. This chapter introduces the current therapies for tissue substrates and highlights the beginning of the tissue engineering field, which leads to the emergence of the three-dimensional printing concept.

1.1 Organ Shortage

The huge success in organ transplantation has led to an increase in demand for transplantable solid organs. Despite the fact that the numbers of organ donors and transplantations have doubled during the last two decades, the demand still far outstrips the supply, leading to a worldwide organ shortage [1]. For instance, in the United States, there were over 97,000 patients on waiting lists requiring organ transplantation in 2007, whereas, only about 29,000 solid organs were transplanted in 2006 [2]. In addition, the increased life expectancy of 7.75 years further leads to long waiting lists. This increase in life expectancy largely results from a reduction in the burden of cardiovascular diseases [3].

Singapore has also severely suffered from a critical organ shortage [4]. In 2007, there were 563 patients waiting for transplantable kidneys. The statistics show a steady climbing trend for new end-stage renal failure. The number of failure cases diagnosed per year has increased from 250 in 1991 to 564 in 1998 and 675 in 2003 [5]. The current average waiting time for a kidney transplant is 8.9 years [4]. However, it is believed that the waiting time is going to be longer in the future due to the increasingly high demand for kidney transplants as well as the median number of deceased organ donors remaining low over the last 5 years. Moreover, the actualisation rate of 30.2% is lower than that of the United States [4]. It is reported that there were 86 potential donors referred to the transplant coordinators of the National Organ Transplant Unit, of which, however, only 26 were eventually actualised as organ donors.

1.2 Current Therapies for Tissue Substitutes

The aim of using tissue substitutes is to replace the diseased tissue. There are two popular therapies, which are organ transplantation and use of medical devices.

1.2.1 *Organ transplantation*

Three dominant clinical applications for organ transplantation are autografting, allografting and xenografting (see Fig. 1.1).

- Autografting

It is defined as tissue that is transplanted from one part of a body to another part in the same individual. Typical examples include skin graft [6], bone graft [7] and anterior cruciate ligament reconstruction with

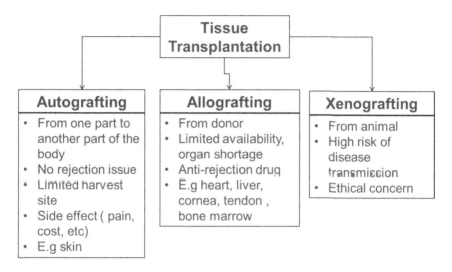

Fig. 1.1. Clinical applications in organ transplantation.

patellar tendon autograft [8]. In general, autografting is the ideal method as it does not induce rejection of the implanted organ and the best clinical results can be obtained. On the other hand, autografting is restricted by some unsolved challenges, such as highly limited available tissues, high surgical costs and potential complications.

• Allografting

Allografting uses tissue/organ (e.g. heart, lungs, liver, kidneys, bone marrow, cartilage and tendons) that is donated by either living or deceased donors. With the continuous improvements of anti-rejection drugs, the allografting method has been widely used in the past decades. Nevertheless, similar to autografting, allografting is subjected to donor shortage. Moreover, relatively high rejection complications together with life-long medication have restricted the application areas of allografting.

- Xenografting

In xenografting, tissue/organ is taken from animal sources and transplanted into a human body. Xenografting could induce more severe problems in tissue rejection and disease transmission. Apart from those technical problems, close attention and concerns have arisen from the use of animals. Xenografting significantly broadens the supply of readily available tissue sources and provides standardised products. With the advances of DNA science, production of transgenic animals which contains organs functionally similar to human will be possible in the future.

1.2.2 *Medical device*

As defined by Food and Drug Administration (FDA), a medical device is 'an instrument, apparatus, implement, machine, contrivance, implant, *in vitro* reagent, or other similar or related article, including a component part, or accessory', which

- is recognised in either the US Pharmacopoeia or National Formulary.
- is intended to be used in diagnosis, cure or prevention of disease, for human or animals, or
- will have significant impact on the body structure or function of human or animals. In addition, the intended purposes of the device cannot be achieved through chemical action or metabolism within the body of human or animals.

Hip and knee implants and artificial heart valves are typical examples of medical devices. In general, medical implants are referred to tissues or devices that are intentionally placed inside or on the surface of the body on purpose, typically for replacing damaged or missing parts of the body e.g. knee implants in prosthetics. Other implants are designed to monitor body functions and deliver necessary medication to the specific tissues and/or organs. A wide range of material resources, such as bone, skin, plastic and metal, has been utilised to fabricate implants depending on application areas.

Implants can either stay inside the body temporarily or permanently. For instance, damaged bones are usually repaired by the use of chemotherapy screws which will subsequently be taken off from the body when they are no longer needed. On the other hand, hip implants are designed to function in the body permanently.

However, certain surgical risks should be addressed. There could be unexpected risks during implant placement and removal. In addition, implant failure and induced infection to the body can be very harmful for the recipient. Some patients can occasionally have reactions to the implants due to improper material used. Other potential complications include bruising, swelling, redness and even infection due to skin contamination during surgery. Additional care and medication will be required when complications occur, which, in the worst case scenario, would require the removal of the implant.

1.3 Tissue Engineering

1.3.1 *Definition*

Tissue engineering is an interdisciplinary field that applies the principles of engineering and the life sciences toward the development of biological substitutes that restore, maintain, or improve tissue function [9]. The term "tissue engineering" was first used in the mid-1980s to describe techniques that were used in manipulation of tissues and organs in surgery. Broadly speaking, TE was also referred to the applications which involved the use of biomaterials and prosthetic devices [10]. In 1987, the concept of TE was formally introduced in medicine and defined as "Tissue Engineering is the application of the principles and methods of engineering and life sciences toward the fundamental understanding of structure-function relationships in normal and pathologic mammalian tissue and the development of biological substitutes to restore, maintain, or improve function" [11].

1.3.2 *Historical development*

Reconstructive surgery techniques aim to improve life quality by replacing missing organ function through rebuilding body structures. Transplantation is a revolutionary and lifesaving technique, which can be considered as the most extreme form of reconstructive surgery transferring tissue and organs from the donor into the patient.

In spite of great success, severe constraints still exist, hindering the development of transplantation. The major issue is the access of sufficient tissue and organs for all the patients. As presented above, many patients will possibly die waiting for available organs due to long transplant waiting lists. In addition, chronic rejection and destruction produced by the immune system over time can create an imbalance of immune surveillance from immunosuppression, which will be highly likely to cause new tumour formation. Moreover, implantable foreign body materials can also cause dislodgment, fracture, migration and infection. Biological changes take place as a result of the abnormal tissue interactions at the foreign body after the tissue is moved from one position to the new position. For example, fatal colon cancers may occur 20 to 30 years later after diverting urine into the colon. These constraints have stimulated a need to seek and develop innovative methods to provide required tissue.

Thus, within this context, TE technology has emerged. New and functional living tissue is produced using existing living cells, that are typically associated with a scaffold or matrix to guide the development of the tissue [11]. A number of new sources of cells such as stem cells have been continuously identified, which draws new interests in the field of TE. Actually, a new term i.e. regenerative medicine was introduced due to the emergence and development of stem cell biology. Scaffolds can either be in a natural form or man-made or even a combination of both. Cells can be cultured within the scaffold and associated with the matrix prior to implantation. Alternatively, they can migrate into the implant after implantation. Cells can be isolated as entirely differentiated cells of the tissue, or manipulated to generate the desired function whilst they are isolated from other tissues [12]. Therefore, the application of TE to human health care can be viewed as a refinement of previously defined principles of medicine.

In the early stage of development, two approaches were widely adopted in TE, namely, cell and tissue culture. In the 1970s, a paediatric orthopaedic surgeon named W. Green undertook a series of experiments to generate cartilage using a chondrocyte culture technique. The chondrocytes were seeded onto bone spicules and implanted in nude mice. Although the experiments ended up unsuccessful, his novel concept to connect and coax cells with appropriately configured scaffolds was very suggestive for the TE development. A few years later, two scientists, Iannas Yannas from M.I.T. and John Burke from Massachusetts General Hospital, conducted a collaborative work on skin generation by seeding and culturing dermal keratinocytes and fibroblasts on protein scaffolds. The generated skin was used to regenerate burn wounds. Other pronounced work includes the seeding of collagen gels with fibroblasts by E. Bell and transferring sheets of keratinocytes onto burn wounds by H. Green [9].

The key point in the history of TE development was given in the Science paper by Joseph Vacanti and Robert Langer [9]. The paper presented the new technology, which may be considered as the birth of tissue engineering. They proposed a concept to design appropriate scaffoldings to facilitate cell delivery and proliferation. Since that time, a large number of research institutions and centres in the US, Europe and Asia have continuously made significant efforts towards this field. TE was catapulted to the public awareness through a British Broadcasting Corporation programme, which explored the potential of tissue-engineered cartilage. By the mid-1990s, TE research was springing up in most of the developed countries and since then, TE has been considered as one of the most promising biomedical technologies of the century [13].

1.3.3 *The promises of tissue engineering*

Tissue engineering is essentially an engineering solution to a medical problem. However, its multidisciplinary nature requires collaboration of cell biologists, materials scientists and mechanical engineers. Basic sciences such as biology, chemistry and physics, and regenerative medicine can benefit from the advances of tissue engineering. The ultimate goal of TE is to produce functional tissues which is able to maintain or restore damaged tissues and/or organs [9]. In general, TE aims to recapitulate some aspects of natural tissue development into functional tissue assembly in terms of cell organisation and differentiation. The use of autologous cells together with biodegradable porous scaffolds and growth factors enable implantable constructs to be specifically engineered and fabricated.

Significant effort and progress have been made in TE in the past 20 years and the pronounced milestones include development of biodegradable scaffolds, TE constructs comprised of cells and biomaterials and bioreactors [14]. The advances in TE have stimulated and led to great success in clinical applications.

Bio-degradable biomaterials are used in every tissue engineered scaffold. The material properties of the scaffold can be customised and further optimised to obtain the desired properties (e.g. mechanical, physical and biological) according to specific microenvironment requirements [15]. The three-dimensional (3D) micro-scale methods currently being investigated are expected to more closely and accurately mimic native tissues [16].

Another promising TE application is to generate vascularised tissues with clinically relevant thicknesses where blood perfusion is established. Some approaches that have shown great potential to achieve this goal are seeding endothelial cells into a scaffold; controlling angiogenic factors release [17]; and micro-fabricating vasculature into a tissue [18]. These methods can also be combined and used to allow 3D vasculature assembly [19].

1.4 Scaffolds in Tissue Engineering

1.4.1 *Process*

The scaffold-based approach is an important concept in tissue engineering. A scaffold is generally referred to a highly porous three-dimensional substrate. Cells that are donated and taken from the patient himself/herself are significantly expanded in culture and then transferred to the scaffold, which provides a surface where cells adhere, proliferate and generate essential elements that make up living tissue (such as the ECM of structural and functional saccharides and proteins). The behaviour and health of the cells seeded inside the scaffold are controlled by not only the scaffold material, but also the internal architecture such as dimensions of the pores, walls, struts and channels.

Within a human body, cells must be very near to the capillary network in order to obtain oxygen and nutrient supplies. For the same reason, cells within a tissue scaffold must be supplied with oxygen and nutrients to maintain life. This is initially achieved by using a highly porous open structure to enable the uninterrupted flow and culture media to access the scaffold in a bioreactor. The engineering component of TE lies in the scaffold design and manufacture [20]. Traditionally, porous scaffolds are produced by a series of routes that lead to a foam-like internal structure with a limited control of scale and a random architecture. Since the 1980s, the rapid prototyping techniques have been continuously advancing, which enables the manufacture of complex and fine-scale internal porous structures.

Rapid prototyping, now known as additive manufacturing (AM), produces complex parts directly from a 3D design file by slicing the object's shape into a number of parallel slices. The shape is manufactured by printing these slices layer-by-layer in a bottom-up manner to build up the structure. For creating solid layers, a range of manufacturing techniques can be of use, including selective laser sintering and melting, selective polymerisation and 3D inkjet printing. Early scaffolds were created from a single biocompatible material. It is now feasible to engineer material consisting of biomimetic components to control the cellular environment [21, 22].

1.4.2 *Example – tissue engineered bladder scaffolds for cystoplasty*

Certain injuries can result in damage of the bladder, which may require repairing or even replacing the injured organ [23]. If drug treatment turns out to be not effective, patients will usually be treated with cystoplasty where donor tissues are needed. Gastrointestinal segments are normally used as donors, which, however, can induce many complications e.g. urolithiasis in the urinary tract [24, 25].

In order to avoid these complications, Atala *et al.* [26] investigated a scaffold-based method to reconstruct bladders utilising autologous bladder tissues for patients with poorly compliant or high-pressure bladders. Based on the bladder biopsy result, a biodegradable bladder-shape scaffold, namely bladder mould made of collagen and polyglycolic acid (GPA), is designed. The volume of the bladder construct is estimated by conducting morphometric analysis. In addition, the patient's age and the pelvic cavity dimensions will also be taken into account. A fabricated scaffold is shown in Fig. 1.2 where muscle and urothelial cells were seeded. The bladder construct was implanted for reconstruction.

Fig. 1.2. The engineered bladder scaffold seeded with cells [26] (copyright 2006, with permission of Elsevier).

1.4.3 *Challenges*

Despite the fact that scaffold-based approach has played an increasingly important role in tissue engineering, there are some limitations that are needed to be overcome in the future.

• Material processability: scaffolds are normally produced by using additive manufacturing (AM) techniques (will be introduced in Chapter 2), which requires input raw material in a very specific form e.g. powder and filament. Thus, the scaffold materials must be compatible with the AM process to be used.

• Mechanical strength of scaffolds: scaffolds should be able to provide desired mechanical properties. Adhesion substrates can neither be too rigid, non-deformable nor compliant, which can be detrimental for signal delivery, cell anchorage and viability [27, 28].

• Degradation rate and product: the degradation and absorption mechanism is significantly affected by a number of factors. Balancing these factors properly has become one of the major challenges [29]. In addition, the scaffold's acidic byproduct effect which is degradation rate-dependent requires further investigation.

• Scaffold morphology: due to edges and grooves on scaffolds, cell adhesion and migration may be influenced by these discontinuities [30].

• Surface topography: cell-matrix interactions are largely influenced by the surface quality of the scaffold [31]. An ideal surface should be reasonably rough but not too rough. This is because, rough surface can enhance cell adhesion, whereas, distinct focal adhesion plaques cannot be obtained if the surface is too rough.

1.4.4 *The promises of bioprinting*

The problems of current tissue engineering technology can be summarised as follows:

• Manual manipulation, not automated or robotic
• Laboratory scale, not industry scale
• 2D simple tissues, not 3D complex organs
• Lack of ordered tissue microstructure and lack of adequate strength

Bioprinting, also known as organ printing (to be presented in detail in Chapter 3), using additive layer-by-layer fabrication manner, is an emerging and fast developing technology. It has shown the potential to circumvent the above limitations in traditional solid scaffold-based TE approach. The ultimate goal of bioprinting is to create three-dimensional vascularised living human organs that are functional and suitable for clinical implantation. The major advantages of bioprinting technology include:

- Scalable reproducible mass production of tissue engineered products;
- Accurate 3D positioning of different types of cells simultaneously achievable;
- Tissues with a high cell density level can be printed and cultured;
- Thick tissue constructs can be vascularised;
- *In situ* printing/dispense of cells

References

[1]   P. J. Hauptman and K. J. O'Connor, "Procurement and allocation of solid organs for transplantation," *New England Journal of Medicine,* vol. 336, pp. 422-431, 1997.

[2]   R. Steinbrook, "Organ donation after cardiac death," *New England Journal of Medicine,* vol. 357, pp. 209-213, 2007.

[3]   W. H. Organization, *Preventing chronic diseases: a vital investment.* Geneva, Switzerland: World Health Organization Press, 2005.

[4]   T. K. Kwek, T. W. Lew, H. L. Tan, and S. Kong, "The transplantable organ shortage in Singapore: has implementation of presumed consent to organ donation made a difference," *Annals Academy of Medicine Singapore,* vol. 38, pp. 346-348, 2009.

[5]   A. Vathsala, "Twenty-five facts about kidney disease in Singapore: in remembrance of World Kidney Day," *Annals Academy of Medicine Singapore,* vol. 36, pp. 157-160, 2007.

[6]   S. T. Boyce, M. J. Goretsky, D. G. Greenhalgh, R. J. Kagan, M. T. Rieman, and G. D. Warden, "Comparative assessment of cultured skin substitutes and native skin autograft for treatment of full-thickness burns," *Annals of surgery,* vol. 222, p. 743, 1995.

[7]   M. J. Yaszemski, R. G. Payne, W. C. Hayes, R. Langer, and A. G. Mikos, "Evolution of bone transplantation: molecular, cellular and tissue strategies to engineer human bone," *Biomaterials,* vol. 17, pp. 175-185, 1996.

[8]   L. J. Salmon, V. J. Russell, K. Refshauge, D. Kader, C. Connolly, J. Linklater, *et al.,* "Long-term Outcome of Endoscopic Anterior Cruciate Ligament Reconstruction With Patellar Tendon Autograft Minimum 13-Year Review," *The American journal of sports medicine,* vol. 34, pp. 721-732, 2006.

[9]   L. Robert and J. P. Vacanti, "Tissue engineering," *Science,* vol. 260, pp. 920-941, 1993.

[10]   R. Skalak and C. Fox, *Tissue Engineering.* Liss, New York, 1988.

[11]   R. Lanza, R. Langer, and J. Vacanti, *Principles of Tissue Engineering.* Burlington, MA, USA: Elsevier, 2007.

[12]   C. A. Vacanti, "History of tissue engineering and a glimpse into its future," *Tissue engineering,* vol. 12, pp. 1137-1142, 2006.

[13]   R. M. Nerem, "Cellular engineering," *Annals of biomedical engineering,* vol. 19, pp. 529-545, 1991.

[14]   M. Papadaki, "Cellular/tissue engineering-promises and challenges in tissue engineering," *Engineering in Medicine and Biology Magazine, IEEE,* vol. 20, pp. 117-126, 2001.

[15]   J. Lahann, S. Mitragotri, T.-N. Tran, H. Kaido, J. Sundaram, I. S. Choi, *et al.,* "A reversibly switching surface," *Science,* vol. 299, pp. 371-374, 2003.

[16]   A. Khademhosseini, R. Langer, J. Borenstein, and J. P. Vacanti, "Microscale technologies for tissue engineering and biology," *Proceedings of the National Academy of Sciences of the United States of America,* vol. 103, pp. 2480-2487, 2006.

[17]   T. P. Richardson, M. C. Peters, A. B. Ennett, and D. J. Mooney, "Polymeric system for dual growth factor delivery," *Nature biotechnology,* vol. 19, pp. 1029-1034, 2001.

[18]   J. T. Borenstein, H. Terai, K. R. King, E. Weinberg, M. Kaazempur-Mofrad, and J. Vacanti, "Microfabrication technology for vascularized tissue engineering," *Biomedical Microdevices,* vol. 4, pp. 167-175, 2002.

[19]   L. Niklason, J. Gao, W. Abbott, K. Hirschi, S. Houser, R. Marini, *et al.,* "Functional arteries grown in vitro," *Science,* vol. 284, pp. 489-493, 1999.

[20]   D. W. Hutmacher, "Scaffolds in tissue engineering bone and cartilage," *Biomaterials,* vol. 21, pp. 2529-2543, 2000.

[21]   M. Lutolf and J. Hubbell, "Synthetic biomaterials as instructive extracellular microenvironments for morphogenesis in tissue engineering," *Nature biotechnology,* vol. 23, pp. 47-55, 2005.

[22]   F. Rosso, G. Marino, A. Giordano, M. Barbarisi, D. Parmeggiani, and A. Barbarisi, "Smart materials as scaffolds for tissue engineering," *Journal of cellular physiology,* vol. 203, pp. 465-470, 2005.

[23]   W. T. Snodgrass and R. Adams, "Initial urologic management of myelomeningocele," *Urologic Clinics of North America,* vol. 31, pp. 427-434, 2004.

[24]   W. McDougal, "Metabolic complications of urinary intestinal diversion," *J Urol,* vol. 147, pp. 1199-1208, 1992.

[25]   T. M. Soergel, M. P. Cain, R. Misseri, T. A. Gardner, M. O. Koch, and R. C. Rink, "Transitional cell carcinoma of the bladder following augmentation cystoplasty for the neuropathic bladder," *The Journal of urology,* vol. 172, pp. 1649-1652, 2004.

[26]   A. Atala, S. B. Bauer, S. Soker, J. J. Yoo, and A. B. Retik, "Tissue-engineered autologous bladders for patients needing cystoplasty," *The Lancet,* vol. 367, pp. 1241-1246, 2006.

[27]   N. Wang, K. Naruse, D. Stamenović, J. J. Fredberg, S. M. Mijailovich, I. M. Tolić Nørrelykke, *et al.,* "Mechanical behavior in living cells consistent with the tensegrity model," *Proceedings of the National Academy of Sciences,* vol. 98, pp. 7765-7770, 2001.

[28]   F. Berthiaume, T. J. Maguire, and M. L. Yarmush, "Tissue engineering and regenerative medicine: history, progress, and challenges," *Annual review of chemical and biomolecular engineering,* vol. 2, pp. 403-430, 2011.

[29]   H. J. Sung, C. Meredith, C. Johnson, and Z. S. Galis, "The effect of scaffold degradation rate on three-dimensional cell growth and angiogenesis," *Biomaterials,* vol. 25, pp. 5735-5742, 2004.

[30]   L. Yin, H. Bien, and E. Entcheva, "Scaffold topography alters intracellular calcium dynamics in cultured cardiomyocyte networks," *American Journal of Physiology-Heart and Circulatory Physiology,* vol. 287, pp. H1276-H1285, 2004.

[31]   W. Y. Yeong, C. K. Chua, K. F. Leong, and M. Chandrasekaran, "Rapid prototyping in tissue engineering: challenges and potential," *Trends in Biotechnology*, vol. 22, pp. 643-652, 2004.

Problems

1.   What are the current therapies for tissue substitutes?
2.   What are the three dominant clinical applications for organ transplantation?
3.   What are the advantages of disadvantages of these three clinical applications?
4.   What is the definition of medical device?
5.   What is tissue engineering? What are the major application areas of tissue engineering?
6.   What are the current limitations of tissue engineering?
7.   What are scaffolds used for in tissue engineering?
8.   What are the limitations and challenges of scaffold-based tissue engineering approach?
9.   What are the advantages of bioprinting as compared with scaffold-based tissue engineering approach?

Chapter 2

Scaffolds for Tissue Engineering

With the ever increasing demand for suitable replacements and organ transplantation, tissue engineering (TE) has become a feasible solution, bringing great hope to patients who are desperate to look for tissue and organ substitutes [1]. In order to obtain substitutes, a popular approach is to fabricate a three-dimensional (3D), biocompatible, biodegradable and porous scaffold functioning as a temporary 3D template for tissue ingrowth [2-8]. Biocompatible polymers, peptides, proteins and hydrogels are the common raw materials for producing scaffolds [9]. In addition to material properties, the scaffold fabrication techniques can largely determine the final characteristics of the scaffold such as porosity and mechanical strength. This chapter describes a number of established processing and fabrication methods including conventional and additive manufacturing (AM) methods. The advantages and disadvantages of the scaffold-based method are also discussed.

2.1 Requirements and Considerations for Fabrication of Scaffolds

A number of techniques for scaffold fabrication have been developed over the past 40 years. Raw material, usually polymers, is handled and shaped into different structures depending on specific TE applications [10, 11]. The fabricated scaffold with desired characteristics such as mechanical strength and surface chemistry directs tissue regeneration [12]. These characteristics can be modified and customised by selecting appropriate material, scaffold components and more importantly, the fabrication technique.

The top priority consideration for all TE applications is always the patient's safety. Therefore, the scaffold must be finally degraded and what is more, the degradation products have to also be biocompatible. It is also crucial that the selected processing method does not have negative impacts on the scaffold in terms of biocompatibility and biodegradability. The scaffold should be able to degrade strictly following a specific time scale, which allows the new growing tissue to gradually replace the scaffold. The scaffold should perform two major functions, which are (1) enabling and directing cell growth within it before implantation; and (2) guiding cell migration into the defect. The scaffold should also contain surface chemistry since this may encourage cell attachment and proliferation of cells. Porous structures are essential for cell adhesion, transport of nutrients and metabolic wastes, confluent tissue formation and sufficient vascularisation of new tissue [13].

In respect of the mechanical properties of a scaffold, they are determined by scaffold geometry, inherent properties of bulk material and the fabrication technique [14]. For instance, polymers with higher crystallinity normally show higher tensile strength. If the crystallinity of polymer chains is reduced due to the processing method used, the resulting scaffold strength is diminished and the scaffold's lifetime is also reduced. Both natural (e.g. collagens, fibrin, carbohydrates and gelatins) and synthetic polymers [e.g. poly(L-lactic acid) (PLLA) and poly(glycolic acid) (PGA)] have been used to manufacture scaffolds. In addition, inorganic materials such as hydroxyapatite have also been utilised in TE scaffolds.

Another primary consideration in scaffold design is the inclusion of bioactive molecules including DNA, proteins and extracellular matrix (ECM)-like peptides. It should be noted that if bioactive molecules are included, the fabrication techniques that activate with those relevant molecules cannot be used since this will affect cell adhesion, signalling and drug and gene delivery. Cell migration, proliferation and differentiation can be significantly promoted through local drug and gene delivery, which improves the quality of tissue regeneration [15].

In the selection of scaffold fabrication techniques, material properties (such as bulk and surface properties) as well as the function that the scaffold is expected to perform should be considered. In addition, the time and cost for fabricating scaffolds relating to the viability of the treatment for the specific patient should also be taken into account. Most fabrication techniques involve applying pressure and/or heat to the material, or organically dissolving it followed by a moulding process to obtain the desired final scaffold geometry. However, harsh conditions are sometimes induced during certain fabrication processes, which are detrimental for cells and bioactive molecules. Therefore, these harsh conditions should be reduced. In fact, each fabrication technique has both distinct benefits and drawbacks, and hence, the selection of an appropriate method should be based on the requirements of the specific type of tissue.

2.2 Conventional Fabrication Techniques of Scaffolds

2.2.1 *Fibre bonding*

Polymer fibres are a viable material for scaffolding because they exhibit outstanding surface-area-to-volume ratio for cell attachment. The earliest TE scaffolds reported were fibre mesh with poor mechanical integrity used for organ regeneration [16]. To overcome this issue, fibre-bonding techniques (see Fig. 2.1) were thus developed, which firmly bind the fibres together at intersection points. The first example of fibre-bonded scaffolds were made of PLLA and PGA [17]. PGA fibres are arranged and placed in a nonwoven mesh. When the temperature increases above the material melting temperature, the fibres bond at their contact points. PGA fibres are encapsulated before heat treatment in order to prevent potential collapses of the melted polymer. PLLA is dissolved in methylene chloride and then cast and dried on the meshed fibres, by which a composite matrix of PGA-PLLA is obtained. An alternative method is to rotate a nonwoven PGA fibre mesh when it is being sprayed with an atomised PLGA or PLLA solution [18]. The polymer solution thus builds up on the PGA fibres and bonds them together. This technique shows its advantages in the fabrication of tubular structures, however, it is unable to fabricate complex 3D structures.

The major advantages of the fibre-bonding fabrication technique are simplicity and retention of the original properties of PGA fibres. The disadvantages are as follows: difficult to control pore size and porosity, limited availability of solvents, and immiscibility of the two different polymers in the melted state.

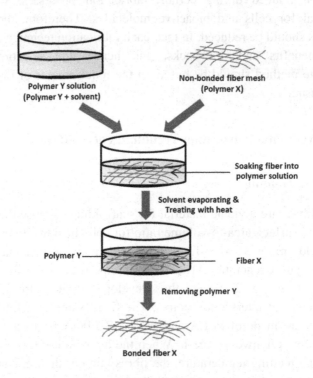

Fig. 2.1. Fibre bonding technique.

2.2.2 *Melt moulding*

The melt moulding technique includes two stages. Firstly, the polymer and porogen particles are mixed and combined in a mould. This is followed by rapidly heating them to the required temperature, i.e. melting temperature (T_m) for semicrystalline polymers and glass transition temperature (T_g) for amorphous polymers. Subsequently, the composite material is taken out from the mould and put into a liquid tank for leaching out the porogen. As a result, the external shape of the produced scaffold is exactly the same as that of the mould. This technique has been used to form PLGA/gelatin microparticle composites with gelatin leaching in deionised water [19]. Virtually any desired geometry of scaffolds can be fabricated by modifying the geometry of the mould in terms of its size and shape. In this technique, porogen particles are used to generate pores and hence, the pore size and the porosity of the scaffold can be controlled through changing the size and the total number of porogen particles respectively. More materials such as hydroxyapatite fibres can be incorporated, providing a bioactive surface and additional mechanical support for cells. Moreover, bioactive molecule delivery can benefit from the use of the mould because this avoids the exposure of the material to harsh organic solvents. However, it should be noted that excessively high moulding temperatures could result in the degradation and inactivation of molecules. Moreover, it is difficult to ensure a consistent interconnectivity of pores, which may limit cell infiltration depth.

2.2.3 *Extrusion*

Traditionally, extrusion is a well-established technique for processing polymers in the industry. In this subsection, the extrusion is a relatively new method for the fabrication of biocompatible porous scaffolds. The first application of extrusion of polymers for TE was to produce tubular scaffolds made of PLLA and PLGA for peripheral nerve regeneration [20]. The polymers were manufactured into a number of membranes with fine thicknesses, which were then tailored to different sizes and loaded into a tool for extrusion. The tool applies pressure and heat to the material, pushing it to go through a die. Finally, the material is pushed out from the nozzle orifice to form cylindrical conduits. After the conduits cool down, they are dipped in water and vacuum dried. This step is to leach the salt. However, the high temperatures during extrusion can induce negative impact on crystallinity, scaffold porosity, and the activity of biomolecules. The extrusion technique is applicable to most biocompatible polymers such as PLGA and PCL. However, this technique can only be used to produce simple shapes, and not interconnected porous structures, hence limiting its application to straight tubular scaffolds.

2.2.4 *Electrospining*

Electrospinning, or electrostatic fibre spinning, shown in Fig. 2.2, is a technique developed based on electrostatic spraying. This method can be used to create scaffolds of biodegradable nonwoven and ultrafine fibres. PCL, PGA and PLGA scaffolds of nanofibres with the porosities of greater than 90% can be electrospun [21]. Raw material, usually polymers, is dissolved in a solvent, which is then loaded into a syringe and ejected onto a grounded collecting surface by charging a high voltage to the capillary [22]. Limitations of this technique are slow production rate, use of organic solvent and poor consistency.

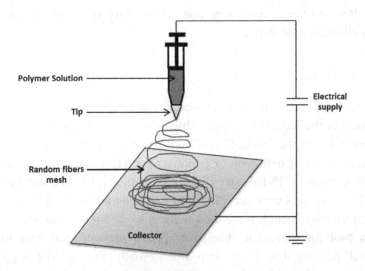

Fig. 2.2. Electrospinning technique.

2.2.5 *Solvent casting and particulate leaching (SCPL)*

With the aim of enhancing control over pore diameter and scaffold porosity, SCPL was developed to generate scaffolds with interconnected porous networks. This technique enables porous scaffolds to be fabricated with specific pore size, porosity, crystallinity and surface-area-to-volume ratio. A polymer is dissolved around a suitable porogen followed by a number of sequential processing steps including casting, drying, solidifying and leaching out the porogen. This process is depicted in Fig. 2.3. At the early development stage of this technique, PLGA and PLLA were utilised as the scaffold material and sieved salt was used as a porogen [23]. The composite material is first heated to a temperature above the T_m and subsequently subjected to an annealing process, which aims to adjust the crystallinity. The porogen is then leached out and a porous PLLA membrane is thus obtained. Other biocompatible porogens can also be used such as lipids [24] and sugars [25]. A solvent exchange system was developed, which utilised the second organic phase to dissolve the porogen whilst not dissolving the polymer. This eliminated the leaching step and as a result, the production time was reduced.

SCPL is able to fabricate scaffolds with controlled pore size (up to 500 μm), porosity (up to 93%) and crystallinity. The major advantage of SCPL is low material consumption; the required amount of polymer to produce a scaffold is relatively small. A typical application is where poly(ethylene glycol) (PEG) and PLGA blends are utilised to fabricate porous foams that are more suitable and less brittle for soft-tissue regeneration [26].

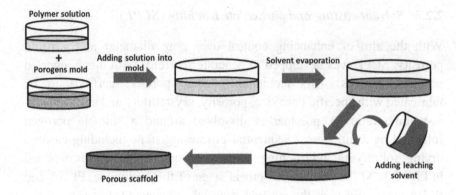

Fig. 2.3. Solvent casting and particulate leaching technique.

2.2.6 *Membrane lamination*

The aim of membrane lamination technique is to create a 3D contour plot of the scaffold shape. A precise contour plot of 3D anatomic geometries are first prepared [27]. Thin layers of porous PLGA or PLLA membranes are produced using the SCPL technique [28]. The neighbouring membranes are chemically bonded with each other by coating a small amount of chloroform solvent at the contacting surfaces. More layers are stacked and bonded until the final 3D structure is complete. This fabrication technique is used to manufacture scaffolds for hard tissues like bone and cartilage with shape-dependent function. It has also been utilised to prepare degradable tubular stents [29].

2.2.7 *Freeze-drying*

Freeze-drying is a rapid fabrication technique for making scaffolds with controllable pore size and porosity. As illustrated in Fig. 2.4, dissolved polymer contained in an organic solution is combined with water and emulsified until homogeneity is achieved [30]. The emulsion is decanted into a metal mould and quickly frozen using liquid nitrogen. The solvent and water are then removed and thus porous scaffolds are formed. Freeze-drying is applicable to fabrication of biocompatible scaffolds made of poly(propylene fumarate) (PPF), PLLA, PGA, PLGA blends [31].

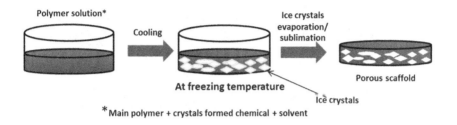

Fig. 2.4. Freeze-drying technique.

2.2.8 *Phase separation*

A biocompatible polymer is dissolved in an appropriate solvent such as dioxane [32], naphthalene [33] and phenol. Bioactive molecules can be added into the solution. The phase separation takes place when reducing the temperature of solution, normally below the T_m of the solvent [34]. By applying liquid nitrogen to the polymer and solvent, they are quenched, thus leading to a two-phase solid. Subsequently, the solidified solvent-rich phase sublimates, resulting in a porous scaffold containing bioactive molecules inside the polymer. The phase separation process is illustrated in Fig. 2.5. It is an ideal candidate for fabricating scaffolds in which the bioactive molecules are embedded since the molecules are not exposed to harsh chemicals or temperatures during production [35]. Nevertheless, the difficulties in incorporating large proteins and acquiring specific drug-release rates are the major drawbacks that restrict the development of this scaffold fabrication technique. The fabrication process is also difficult to control since the scaffold morphology can be significantly affected when changing the process parameters and thermal quenching strategy.

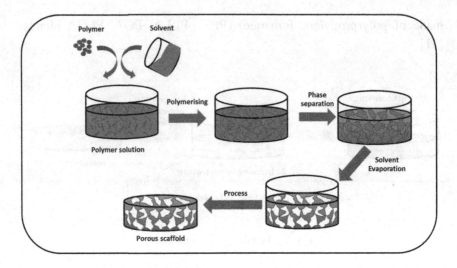

Fig. 2.5. Phase separation technique.

2.2.9 *Gas foaming*

A major concern of using SCPL fabricated scaffolds is inflammatory responses after implantation induced by the remnants of the organic solvents used in the SCPL scaffold fabrication process. Therefore, the gas-foaming fabrication method [36] is suggested if avoiding the use of organic solvents is not required. As shown in Fig. 2.6, compressed polymer disks are processed with high-pressure carbon dioxide. As the pressure reduces, pore formation and nucleation take place in the polymer matrix. However, this technique is only able to fabricate scaffolds with closed-pore morphology [37].

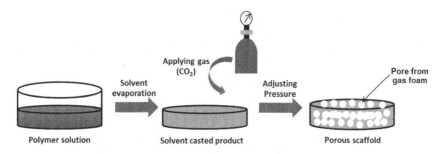

Fig. 2.6. Gas foaming technique.

2.2.10 *Peptide self-assembly*

Peptide nanofibres have recently been used as a synthetic ECM in scaffolds because they are able to self-assemble to form stable scaffolds [38]. Typically, peptides are comprised of ionic, self-complementary sequences with interchanging hydrophilic and hydrophobic domains [39]. It has been found that peptide-based scaffolds are advantageous for culturing chondrocytes, osteoblasts and hepatocytes *in vitro*. The structures of self-assembling peptide allow attached cells to remain in their original 3D geometries. The sizes of aggregate scaffolds can reach the centimetre scale due to the individual fibres being as small as 5 nm [40]. Typically, self-assembling peptides form stable β-sheets in physiological solutions or water.

2.2.11 *Polymer/ceramic composite foam fabrication*

Scaffolds used for bone replacement require high mechanical strength and irregular shapes for most bone defects. Solvent-casting is first employed, dispersing hydroxyapatite (HA) fibres or micro-particles and a porogen into a PLGA chloride solution. The porogen is then leached out from the solution after the solvent gradually evaporates. As a result, the composite scaffold is obtained, which contains both PLGA and HA fibres/micro-particles [41]. Scaffolds fabricated by this technique exhibit superior compressive strength comparable to cancellous bone [42].

2.2.12 *Summary*

Table 2.1 summarises some typical conventional techniques for fabrication of three-dimensional scaffolds.

Table 2.1. Polymer scaffold processing techniques for tissue engineering.

Fabrication process	Advantage	Disadvantage	Ref
Fibre bonding	High porosity	Limited application to other polymers Lack required mechanical strength for the load-bearing tissues Solvent residue may be harmful	[43, 44]
Melt moulding	Independent control of porosity and pore size Macro shape control	High temperature required for nanomorphous polymer	[45-47]
Solvent casting and particulate leaching	Controlled porosity, up to 93% Independent control of porosity and pore size Crystallinity can be tailored	Limited to membranes up to 3mm thick Lack required mechanical strength for the load-bearing tissues	[48-50]
Membrane lamination	3D scaffold	Low mechanical strength	[28, 29]
phase separation	Nondecreased activity of the molecule	Difficult to control precisely scaffold morphology Solvent residue may be harmful	[51, 52]
Polymer/ceramic fibre composite fabrication	Superior compressive strength Independent control of porosity and pore size	Solvent residue may be harmful	[53, 54]

2.3 Additive Manufacturing Techniques of Scaffolds: Direct Methods

Additive Manufacturing (AM) techniques are able to directly create a physical model represented by computer aided design (CAD) data. Every part is generated in a layer-by-layer manner. This section presents the AM techniques that can be used to fabricate scaffolds directly or indirectly.

AM systems such as Selective Laser Sintering (SLS), Colour Jet Printing (CJP) and Fused Deposition Modelling (FDM) [55, 56] have the capability and feasibility to fabricate porous structures for TE applications. In this section, the AM-based tissue scaffold fabrication techniques are classified into two typical processes: the melt-dissolution deposition and the particle bonding techniques.

2.3.1 *Melt-dissolution deposition technique*

In a melt-dissolution deposition process, layers are built and stacked by extruding a strand of melted material through a nozzle while it is moving around the horizontal plane of the layer cross-section. The newly deposited material quickly solidifies and bonds to the previous layer underneath. This process continues until a complex 3D solid object is fully generated.

Porosity in the horizontal plane is created and controlled by changing the infill hatch i.e. the spacing between neighbouring filaments (see Fig. 2.7). The gaps in the vertical direction are created by depositing the subsequent layer at a certain angle in relation to the previous layer. A structure with desired porosity will be generated as a result of repetitive pattern drawing. For example, the interconnected square channels in Fig. 2.7 were produced by using a pattern setting of 0° or 90°. The structure of the scaffold is usually regular and highly reproducible. The machine parameters related to the material properties need to be precisely controlled to reduce filament deflection. Filaments are orthogonally aligned, which creates grooves at the intersection point between successive layers, facilitating cells to span across the grooves to eventually cellularise the entire structure. The most representative melt-dissolution deposition process is FDM.

Fig. 2.7. Structure produced using fused deposition modelling.

3.3.1.1 *Fused deposition modelling*

A filament of material is fed into the liquefier and melted into a semi-liquid state before deposition through a nozzle [57, 58]. This process needs to be operated in a heated chamber in order to maintain sufficient fusion energy.

FDM has been utilised to manufacture functional scaffolds. A typical FDM system is shown in Figure 2.8. Samar *et al.* [59] fabricated a scaffold made of polypropylene-tricalcium phosphate (PP-TCP). The pore size of the scaffold is 160 µm and the mechanical strength is 12.7 MPa, which is higher than the tensile strength of natural cancellous bone (normally 7.4MPa [60]). Zein *et al.* [61] produced PCL scaffolds that had a honeycomb structure with a 160-770 µm channel size. PCL and PCL-HA scaffolds were also manufactured, on which human mesenchymal progenitor cells were seeded, enabling cell proliferation [62].

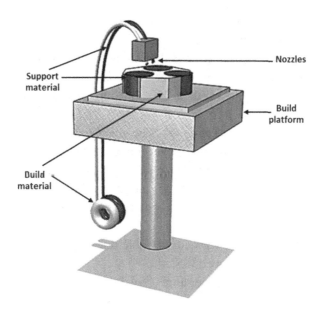

Fig. 2.8. Fused deposition modelling technique.

However, certain limitations should be addressed. FDM requires input material to be of a specific size. Moreover, any changes to material properties require recalibration of the roller feeding parameters. The resolution of the FDM process is 250 μm, which is lower than other AM methods to be introduced in the following sections. Available materials for FDM are quite limited. Natural polymers cannot be used since materials have to be in the filament form and melted during extrusion. The high system operating temperature impedes the incorporation of biomolecules into the scaffold. In addition, the deposited material quickly solidifies into dense filaments, which is detrimental for microporosity formation. This further leads to ineffective neovascularisation and cell attachment [63].

In order to circumvent the above limitations, improvements are continuously made, leading to the emergence of variants of the FDM technique. Lower operating temperatures can be applied and precursor filaments are not necessarily required. These variants include precision extruding deposition (PED) [64], 3D fibre-deposition technique [26] and precise extrusion manufacturing (PEM) [65].

PED: The extruder is equipped with a heating unit to melt the feedstock material. As a result, the need to produce precursor filaments is eliminated. PCL scaffolds with a 250 μm pore size are manufactured [64].

3D fibre-deposition technique: The feedstock material is in a granular or pellet form that can be directly poured into the heated liquefier. Poly(ethyleneglycol)-terephthalate-poly(butylenes terephthalate) (PEGT–PBT) scaffolds are fabricated for articulate TE applications [66]. Material flow is regulated by controlling the pressure in the syringe.

PEM: PLLA scaffolds with controllable pore sizes ranging from 200 to 500 μm have been produced [65].

Due to the elevated temperatures involved in the melting process, scaffold bioactivity may be damaged. New systems have been developed to replace the melting process with material dissolution. Some typical systems are multi-nozzle deposition manufacturing (MDM) [67], low-temperature deposition manufacturing (LDM) [65], robocasting [68] and pressure-assisted microsyringes (PAM) [69].

LDM: The scaffold generation cycle is implemented in a low temperature environment (normally under 0°C) [70]. A PLLA–TCP pipe scaffold is fabricated.

MDM: It is an enhanced version of LDM, with a wider range of available materials. More jetting nozzles are integrated into the system [68]. Support can be built using nontoxic water, facilitating easy removal after deposition. Zhuo *et al.* [67] embedded biomolecules (in the bone morphogenic protein form) in the bulk material, which were then gradually released as the scaffold degraded.

Robocasting: This system can build concentrated and pseudoplastic-like colloidal suspension [68]. A 3D micro-vascular network was created by Therriault *et al.* [71], using fugitive organic ink.

PAM: This method utilises a microsypringe to expel the dissolved polymer under a constant pressure to form the designed pattern. The significant advantage of this method is the high resolution, which can achieve a cellular scale. The PLLA scaffolds with line width of 20 μm were fabricated by Vozzi *et al.* [69], demonstrating that the PAM performance was comparable to that of lithography [72].

2.3.2 *Particle-bonding techniques*

Particle-bonding techniques use powder as the raw material where particles are selectively bonded. A 3D structure is formed through bonding thin 2D layers one upon another. Unprocessed powder supports the scaffold being manufactured. As a result, overhanging features, through channels and holes can be easily fabricated. These techniques are very popular in fabrication porous structures due to both macroporosity and microporosity being controllable. Pore architectures are controlled through manipulating the bonding region. Nevertheless, the pore size is still restricted by the powder size of the raw material. Bigger pores can be created by adding porogen into the powder bed prior to the bonding process.

Topographical cues have been found to have a significant effect on the behaviour of cells [73]. Stretch receptors are activated when cells are attached to the scaffold. Receptors on the surface of the scaffold are subjected to varying deformation degrees, resulting in activation of pathways of cell signal transduction. Hence, cell attachment can benefit from using scaffolds produced by particle-bonding techniques. Representative systems in this class include SLS [74-79], CJP [41] and TheriForm™ [80].

3.3.2.1 *Selective laser sintering*

In a SLS process (see Fig. 2.9), a laser beam is used to selectively scan the surface of the powder according to specific cross-sectional profiles [81-87]. The powder is heated to its glass transition temperature, leading to material deformation and fusion. SLS is suitable for fabrication of porous ceramic matrices for bone implantation [88-90].

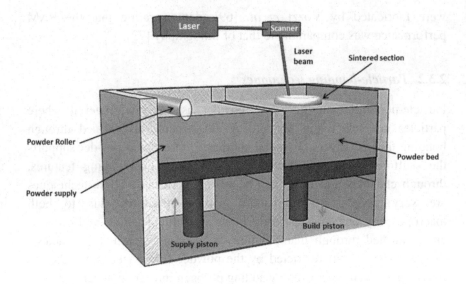

Fig. 2.9. Selective laser sintering technique.

3.3.2.2 *Colour jet printing*

A CJP system is depicted in Fig. 2.10, where adhesive droplets are continuously ejected through an inkjet printhead, by which layers of powder particles are bonded with each other to form a 3D object. This technique has been used to produce PLGA scaffolds of 60% porosity, mixed with an organic solvent and some salt particles [91].

3.3.2.3 *TheriForm™*

The overall configuration and principle of this technique is similar to CJP. Binder droplets are ejected onto the specific regions of the polymer powder. The powder is thus swelled and dissolved [92].

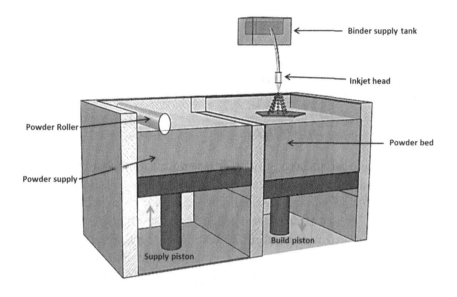

Fig. 2.10. Colour jet printing technique.

2.4 Additive Manufacturing Techniques of Scaffolds: Indirect Methods

Another major application of AM techniques is the manufacture of sacrificial moulds for producing TE scaffolds. Material is first cast in a mould and subsequently removed from it to obtain the final scaffold required. These AM techniques allow the operator to control both internal and external morphology of the scaffold. Moreover, the original material properties are preserved since the scaffold is not subjected to the heating process. Typical techniques include droplet deposition [93], melt deposition [94] and photopolymerisation [95].

2.4.1 *Droplet Deposition*

Droplet deposition originates from inkjet printing technology. Molten thermoplastic droplets are continuously extruded onto a surface. Local melting of the layers is induced by the thermal energy. A typical system is described below.

3.4.1.1 *ModelMakerII™ (MMII)*

This system is equipped with two jets for depositing polymer material for the scaffold and wax-like material for the support [96, 97]. Both scaffold and support materials are melted into a liquid state and stored in reservoirs. A stream of droplets is jetted from the printhead whilst it is moving on the horizontal platform. Upon obtaining a layer, a milling tool is used to machine the layer, ensuring a uniform layer thickness.

Taboas *et al.* [98] fabricated PLLA scaffolds with both macro- and micro-pores for implantation in a trabecular bone defect. Local pores with up to 100 µm wide voids were obtained by using porogen. Materials with a low thermal expansion coefficient, such as calcium phosphate, are available for MMII. Fracture risks during pyrolysis can be significantly reduced as a result of coefficient mismatching with the ceramic. Limpanuphap and Derby [99] used MMII to control porosity during the TCP scaffold building process. It is also reported that HA porous scaffolds with specific macro-architecture where channels are approximately 350–400 µm wide can be generated [100]. Collagen scaffolds and chitosan-collagen scaffolds with intricate internal channels of 135 µm (Fig. 2.11) were fabricated by Sachlos *et al.* [101]. These channels can act as flow channels when they are coupled with specific bioreactors, which improve the culture medium perfusion. Yeong *et al.* [97] used an inkjet MMII to fabricate collagen scaffolds with pore size ranging from 20 µm to 50 µm.

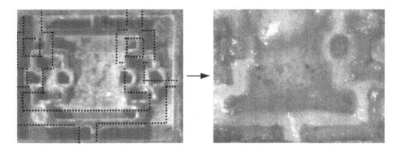

Fig. 2.11. Scaffold (right) produced using ModelMakerIITM-built mould (left).

2.4.2 Melt-dissolution deposition

The process principle of this class of techniques, where FDM is a representative method, is described in the previous sections. Bose *et al.* [94] manufactured β-TCP ceramic and alumina scaffolds. The moulds were made of thermoplastic polymer, and the pore sizes and porosity of the final scaffolds were 300-500 μm and 45% respectively. Since pore size and porosity can be easily controlled and manipulated by changing relevant FDM process parameters, FDM is widely used to explore the mechanical properties and biological responses of the scaffolds with different pore sizes and porosity.

2.4.3 Photopolymerisation

In general, a laser beam is used to intensively irradiate the surface of a liquid photopolymer resin. Due to the thermal and optical energy provided by the laser beam, chemical reaction takes place on the irradiation areas, by which the liquid resin is quickly transformed into a solid state.

3.4.3.1 Stereolithography Apparatus (SLA)

A SLA system is depicted in Figure 2.12, where an ultraviolet (UV) laser selectively scans the surface of photocurable resin. The photopolymerisation thus takes place and scanned resin rapidly solidifies, and the un-scanned areas are still in the liquid phase. The elevator moves downwards a certain distance to ensure the solidified polymer is sunk and fully covered by the liquid resin. The laser then begins to scan the liquid surface, forming the next layer of solid polymer atop the previous one.

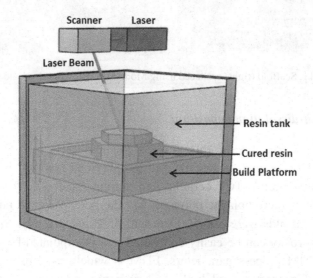

Fig. 2.12. Stereolithography apparatus technique.

Chu *et al.* [95] used this technique to manufacture HA-based porous implants. The crucial step was the casting of HA-acrylate suspension into a mould, by which the scaffold with interconnected channels of 366 μm in resolution was generated. *In vivo* study was also conducted to explore two architecture designs, namely radial and orthogonal channels [102]. It was found that the geometry of the regenerated bone tissue can be controlled by adopting appropriate designs of scaffold architectures.

3.4.3.2 *Two-photon polymerisation*

Two-photon excitation provides energy to activate physical or chemical processes with high resolution in three-dimensional spaces, which facilitates the development of three-dimensional lithographic microfabrication [103, 104]. The two-photon absorption probability depends on intensity and thus, the absorption is constrained at the focus to a volume of order λ^3 under tight-focusing conditions (λ is the laser wavelength) [105].

When tightly focusing a laser beam into the volume of a photosensitive resin, near infrared femtosecond laser pulses are non-linearly absorbed by the material within the focal volume [106]. If the light intensity is sufficient at the focus, the polymerisation process is initiated. By moving the laser focus through the material, three-dimensional patterns with a resolution down to 100 nm can be fabricated in a single setup [107]. The two-photon polymerisation process is illustrated in Figure 2.13. This technique has been used to create hydrogel constructs such as Poly(ethylene glycol) (PEG) hydrogels [108, 109].

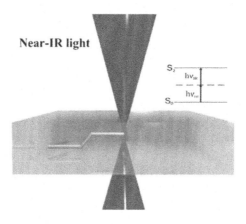

Fig. 2.13. Schematic illustration of two-photon polymerisation process [105] (copyright 2011, with permission of Elsevier).

2.4.4 *Summary*

Table 2.2 outlines a number of AM methods applicable to scaffold fabrication and compares their strengths and weaknesses [5].

Table 2.2. Single lined table captions are centered to the table width. Long captions are justified to the table width manually.

AM system	Resolution (μm)	Material	Advantage	Disadvantage	Ref
Melt-dissolution deposition technique					
Fused deposition modelling	250	PCL[a], PP-TCP, PLA-HA, PCL-TCP	Good mechanical strength; Versatile in lay-down pattern design	High temperature; Need to produce filament material; Narrow processing window; Rigid filament	[110-112]
3D fibre-deposition technique	250	PEGT-PBT	Input material in pellet form; Reduced preparation time	High temperature; Rigid filament	[113]
Precision extruding deposition	250	PCL	Input material in pellet form	High temperature; Rigid filament	[114]
Precise extrusion manufacturing	200-500	PLLA-TCP	Input material in pellet form	High temperature; Rigid filament	[115]
Low-temperature deposition manufacturing	350	PLLA-TCP	Biomolecule can be incorporated	Solvent is used; Freeze drying is required	[116]
Multi-nozzle deposition manufacturing	400	PLLA-TCP	Enhanced range of materials; Biomolecule can be incorporated;	Solvent is used; Freeze drying is required	[117, 118]
Pressure-assisted microsyringe	5-500	PCL, PLLA	Enhanced range of materials; Biomolecule can be incorporated; Very fine resolution achievable	Small nozzle inhibits incorporation of particle; Narrow range of printable viscosities; Solvent is used	[119-121]

Robocasting	100-1000	Organic ink	Enhanced range of materials;	Precise control of ink properties is crucial	[122, 123]
3D bioplotter	250	Hydrogel	Enhanced range of materials; Biomolecule can be incorporated;	Low mechanical strength; Smooth surface; Low accuracy; Slow processing; Precise control of properties of material and medium; Calibration for new material	[124, 125]
Rapid prototyping robotic dispensing system	400-1000	Chitosanchitosan-HA	Enhanced range of materials; Biomolecule can be incorporated;	Precise control properties of material and medium; Freeze drying is required	[126]
Particle bonding techniques					
3-dimensional printing™	200	PLGA, starch-based polymer	Micro-porosity induced in the scaffold; Enhanced range of materials; Water can be used as binder; No support structure needed; Fast processing	Material must be in powder form; Limited mechanical strength; Powdery surface finish; Trapped powder issue; Post-processing is required	[127]
TheriForm™	300	PLLA	Micro-porosity induced in the scaffold; Enhanced range of materials; Non-organic binder is possible; No support structure needed;	Material must be in powder form; Powdery surface finish; Trapped powder issue	[128]

Method		Material	Advantages	Disadvantages	Ref.
Selective laser sintering	450	PEEK-HA, PCL	Fast processing Micro-porosity induced in the scaffold; Enhanced range of materials; No support structure needed; Fast processing	Material must be in powder form; High temperature; Powdery surface finish; Trapped powder issue	[87, 129, 130]
Indirect AM fabrication method					
Melt deposition	250	Thermoplastic polymer	Enhanced range of materials; Control of external and internal morphology	Multi-steps involved	[131]
Droplet deposition	150	Wax	Enhanced range of materials; Control of external and internal morphology	Multi-steps involved	[132]
Photo-polymerisation	366	Resin	Enhanced range of materials; Control of external and internal morphology	Multi-steps involved	[133]

aAbbreviations: HA, hydroxyapatite; PCL, polycaprolactone; PEEK-HA, polyetheretherketone-hydroxyapatite; PEGT-PBT, poly(ethylene glycol)-terephthalate-poly(butylenes terephthalate; PLLA, poly-L-lactide; PP-TCP, polypropylene-tricalcium phosphate

2.5 Applications of Additive Manufactured Scaffolds

Additive manufacturing has been recognised as a viable method for fabrication of scaffolds based on the above mentioned advantages. The process is controllable, precise, patient-specific and easily scalable. The following table shows some AM scaffolds that are commercially available.

Table 2.2. Applications of additive manufactured scaffolds.

Application	Material	Method	Outcome
OsteoFab Patient specific cranial implant [a]	OXPE-KK	Laser sintering	The device is built individually for each patient to correct defects in cranial bone. The device is constructed with the use of the patient's CT imaging data and computer-aided design to determine the dimension of each implant.
Alvelac™ Dental plug [b]	PLGA		Alvelac™ is intended to help maintain the original height and width of the socket so that there would be minimal bone loss and a good bone structure maintained. It is designed to maintain socket height and width, which will allow for natural bone healing. It will be resorbed in approximately two to six months.
Osteopore PCL scaffold Bone Filler [c]	PCL	Fused deposition modelling	Osteopore PCL scaffold Bone Filler are made of filaments of 3D inter-woven bioresorbable polymer. The pores are inter-connected and the polymer has been regarded as cell tissue compatible.
TRS Cranial Bone Void Filler [d]	PCL	Laser sintering	TRS Cranial Bone Void Filler (TRS C-BVF) is a synthetic, porous, osteocondluctive, bone void filler made from POIL polycaprolactone (C6H-1002) X which will degrade and resorb fully in vivo by hydrolysis and is subsequently metabolized by the body, and hydroxylapatite

[a] Oxford Perfomance Materials, LLC.
http://www.accessdata.fda.gov/cdrh_docs/pdf12/K121818.pdf
[b] www.bio-scaffold.com
[c] http://www.osteopore.com.sg/tech.htm
[d] http://tissuesys.com/technology
http://www.accessdata.fda.gov/cdrh_docs/pdf12/K123633.pdf

			(CalO(P04 6.(OH)2) with a calcium phosphate bone mineral coating (Hydroxylapatite and Octacalcium phosphate). TRS BVF has an interconnected porous structure that acts as an osteocondluctive matrix for the ingrowth of bone.

2.6 Challenges of Additive Manufacturing in Tissue Engineering

- AM materials: any AM technique is material dependent, and there is no universal AM process. Materials that are AM-able for TE applications are limited.
- AM design: a typical AM process starts with a CAD model. However, native tissue has a hierarchical structure complicated with numerous bifurcations. Replicating this reality in the CAD model is challenging.
- AM resolution: an ideal scaffold should be an equivalent of an ECM. This requires a fabrication resolution at nanoscale (<100 nm). No AM can fabricate an ECM yet.

2.7 Clinical Considerations with Scaffold-Based Tissue Engineering

An ideal biodegradable polymer scaffold should (1) be nontoxic; (2) be able to maintain mechanical integrity to facilitate tissue growth, integration and differentiation; (3) be capable of controlling degradation rate; (4) be nonimmunogenic; and (5) not cause infection. However, concerns including the long-term safety of scaffold degradation products, immunogenicity, and the risk of infection or transmission of disease remain unsolved.

2.7.1 *Immune reactions*

A serious problem of scaffold-based TE is that scaffolds may cause undesirable immune reactions including acute allergic responses or late-phase responses, and inflammation [134]. Scaffolds are also likely to induce an autoimmune response [135]. Additionally, scaffold degradation byproducts may trigger immune responses. Tiny metal particles shed from the metallic medical implants (e.g. debris from artificial joint prostheses [136]) can induce a specific form of inflammation (i.e. metallosis). It is also highly dangerous if the scaffold degradation byproducts accumulate gradually over years because that will elicit chronic diseases related to inflammatory responses.

2.7.2 *Degradation of Scaffolds in vivo*

Classic biodegradable polymers can be gradually absorbed by environmental bacteria through a process that is different from the physio-logical degradation processes. Biodegradation may result in toxicity in two ways: either a degradation product is metabolised to a toxic (i.e. by liver enzymes) or it is directly toxic.

Most synthesised biodegradable polymers are decomposed by hydrolysis, leading to the accumulation of acids. This is highly dangerous as an excess of acids can alter the pH of the microenvironment and even worse, cause more direct toxicity. Moreover, some scaffolds are destroyed by macrophages, resulting in an inflammatory reaction.

As bone substitute scaffolds can be gradually replaced by real bone, most other biomaterial scaffolds disappear during degradation, which will leave a potential space that can hinder repair. In cartilage repair, for example, when the scaffold is degraded, the space left there may no longer be filled with chondrocytes owing to the low proliferative capacity of the cells. Eventually, these spaces are likely to form tiny cracks, which further deteriorate the smooth cartilage surface.

2.7.3 *Risks of Infection*

The two major sources that are likely to induce infection from implanted scaffolds include: (1) cells and infections that emerge from the bacterial biofilm, which is formed and attached on the surface of the scaffold after implantation; and (2) pathogens transmitted directly from the scaffold.

3.7.3.1 *Biofilms*

The major source of infection is induced by the biofilm formed around the surface of the implanted surface [137]. Medical devices and implants, such as orthopaedic or dental implants and catheters, are now ubiquitous in clinical practice. As a result, the frequency of device-related infections continuously increases [138]. Most of infections result from staphylococci, which generally do not respond to antibiotic therapy. In this case, the implanted device has to be removed. For example, infection induced in artificial joint replacement surgery is a serious complication that cannot be cured by other methods other than removing the implant [139]. Infection is also the most common reason for removal of breast implant [140].

In vivo microbial contamination of medical devices is different from infection of natural tissues. Due to the lack of an immune system or bloodstream in the devices, the microorganisms start to form a bacterial biofilm as long as and as soon as they invade through skin scratches, airways, wounds and attach to the surface of the implanted devices [141]. The biofilm consists of glycoproteins and polysaccharides secreted by microorganisms. Biofilm-protected microorganisms are resistant to host immunity, physical removal and antibiotics. Moreover, long-term antibiotic treatment may result in the increase of the risk of antibiotic resistance due to most antibiotics not being able to diffuse inside the biofilm completely. Furthermore, since biofilms develop slowly, resulting infections may emerge a few years after implantation.

3.7.3.2 *Risks of disease transmission by scaffolds*

There are some animal-derived scaffolds such as collagen gels and amniotic membranes. Animal pathogens induced by implanted scaffolds can cause severe infections in humans and are sometimes lethal. Bovine spongiform encephalopathy is a typical example. In addition, it is highly likely that many animal pathogens exist and remain undiscovered, which may potentially cause human disease or death.

Though scaffolds used are mostly biodegradable, degradation is typically a very slow process which may take a few years. When infection takes places at the scaffold site, the scaffold has to be surgically removed, which disrupts tissue repair and increases the risk of further damage. A large amount of effort has been focused on the development of infection-resistant biomaterials and three representative materials are ceramics that slowly release antibiotics [142], silver ion-coated materials, and antibacterial adhesion polymers [143]. However, these antibacterial factors might also be detrimental for implanted cells. Overall, scaffold-based tissue engineering methods hold great clinical potential but significant safety concerns still remain unsolved. Nevertheless, the risks in a clinical setting can be minimised by appropriate sterilisation.

References

[1]   J. Tan, C. Chua, K. Leong, K. S. Chian, W. Leong, and L. Tan, "Esophageal tissue engineering: An in-depth review on scaffold design," *Biotechnology and bioengineering,* vol. 109, pp. 1-15, 2012.

[2]   D. Ma, F. Lin, and C. Chua, "Rapid prototyping applications in medicine. Part 1: NURBS-based volume modelling," *The International Journal of Advanced Manufacturing Technology,* vol. 18, pp. 103-117, 2001.

[3]   D. Ma, F. Lin, and C. Chua, "Rapid prototyping applications in medicine. Part 2: STL file generation and case studies," *The International Journal of Advanced Manufacturing Technology,* vol. 18, pp. 118-127, 2001.

[4]   S. Yang, K. F. Leong, Z. Du, and C. K. Chua, "The design of scaffolds for use in tissue engineering. Part I. Traditional factors," *Tissue engineering*, vol. 7, pp. 679-689, 2001.

[5]   W. Y. Yeong, C. K. Chua, K. F. Leong, and M. Chandrasekaran, "Rapid prototyping in tissue engineering: challenges and potential," *Trends in Biotechnology*, vol. 22, pp. 643-652, 2004.

[6]   K. Leong, C. Cheah, and C. Chua, "Solid freeform fabrication of three-dimensional scaffolds for engineering replacement tissues and organs," *Biomaterials*, vol. 24, pp. 2363-2378, 2003.

[7]   L. Dan, C. K. Chua, and K. F. Leong, "Fibroblast response to interstitial flow: A state-of-the-art review," *Biotechnology and bioengineering*, vol. 107, pp. 1-10, 2010.

[8]   C. Liew, K. Leong, C. Chua, and Z. Du, "Dual material rapid prototyping techniques for the development of biomedical devices. Part 1: Space creation," *The International Journal of Advanced Manufacturing Technology*, vol. 18, pp. 717-723, 2001.

[9]   L. E. Freed, G. Vunjak Novakovic, R. J. Biron, D. B. Eagles, D. C. Lesnoy, S. K. Barlow, *et al.*, "Biodegradable polymer scaffolds for tissue engineering," *Nature Biotechnology*, vol. 12, pp. 689-693, 1994.

[10]   K. Leong, C. Chua, N. Sudarmadji, and W. Yeong, "Engineering functionally graded tissue engineering scaffolds," *Journal of the mechanical behavior of biomedical materials*, vol. 1, pp. 140-152, 2008.

[11]   M. Naing, C. Chua, K. Leong, and Y. Wang, "Fabrication of customised scaffolds using computer-aided design and rapid prototyping techniques," *Rapid Prototyping Journal*, vol. 11, pp. 249-259, 2005.

[12]   D. W. Hutmacher, "Scaffold design and fabrication technologies for engineering tissues - state of the art and future perspectives," *Journal of Biomaterials Science, Polymer Edition*, vol. 12, pp. 107-124, 2001.

[13]   R. Lanza, R. Langer, and J. Vacanti, *Principles of Tissue Engineering*. Burlington, MA, USA: Elsevier, 2007.

[14]   D. W. Hutmacher, "Scaffolds in tissue engineering bone and cartilage," *Biomaterials*, vol. 21, pp. 2529-2543, 2000.

[15]   J.-H. Jang, T. L. Houchin, and L. D. Shea, "Gene delivery from polymer scaffolds for tissue engineering," *Expert review of medical devices*, vol. 1, pp. 127-138, 2004.

[16] L. Cima, J. Vacanti, C. Vacanti, D. Ingber, D. Mooney, and R. Langer, "Tissue engineering by cell transplantation using degradable polymer substrates," *Journal of biomechanical engineering*, vol. 113, pp. 143-151, 1991.

[17] A. G. Mikos, Y. Bao, L. G. Cima, D. E. Ingber, J. P. Vacanti, and R. Langer, "Preparation of poly (glycolic acid) bonded fiber structures for cell attachment and transplantation," *Journal of biomedical materials research*, vol. 27, pp. 183-189, 1993.

[18] D. Mooney, C. Mazzoni, C. Breuer, K. McNamara, D. Hern, J. Vacanti, *et al.*, "Stabilized polyglycolic acid fibre-based tubes for tissue engineering," *Biomaterials*, vol. 17, pp. 115-124, 1996.

[19] R. C. Thomson, M. J. Yaszemski, J. M. Powers, and A. G. Mikos, "Fabrication of biodegradable polymer scaffolds to engineer trabecular bone," *Journal of Biomaterials Science, Polymer Edition*, vol. 7, pp. 23-38, 1996.

[20] M. S. Widmer, P. K. Gupta, L. Lu, R. K. Meszlenyi, G. R. Evans, K. Brandt, *et al.*, "Manufacture of porous biodegradable polymer conduits by an extrusion process for guided tissue regeneration," *Biomaterials*, vol. 19, pp. 1945-1955, 1998.

[21] H. Yoshimoto, Y. Shin, H. Terai, and J. Vacanti, "A biodegradable nanofiber scaffold by electrospinning and its potential for bone tissue engineering," *Biomaterials*, vol. 24, pp. 2077-2082, 2003.

[22] Q. P. Pham, U. Sharma, and A. G. Mikos, "Electrospinning of polymeric nanofibers for tissue engineering applications: a review," *Tissue engineering*, vol. 12, pp. 1197-1211, 2006.

[23] A. G. Mikos, M. D. Lyman, L. E. Freed, and R. Langer, "Wetting of poly (L-lactic acid) and poly (DL-lactic-co-glycolic acid) foams for tissue culture," *Biomaterials*, vol. 15, pp. 55-58, 1994.

[24] M. Hacker, J. Tessmar, M. Neubauer, A. Blaimer, T. Blunk, A. Göpferich, *et al.*, "Towards biomimetic scaffolds: anhydrous scaffold fabrication from biodegradable amine-reactive diblock copolymers," *Biomaterials*, vol. 24, pp. 4459-4473, 2003.

[25] C. E. Holy, S. M. Dang, J. E. Davies, and M. S. Shoichet, "In vitro degradation of a novel poly (lactide-co-glycolide) 75/25 foam," *Biomaterials*, vol. 20, pp. 1177-1185, 1999.

[26] M. C. Wake, P. K. Gupta, and A. G. Mikos, "Fabrication of pliable biodegradable polymer foams to engineer soft tissues," *Cell transplantation*, vol. 5, pp. 465-473, 1996.

[27]   L. G. Cima, R. S. Langer, A. G. Mikos, G. Sarakinos, and J. P. Vacanti, "Biocompatible polymer membranes and methods of preparation of three dimensional membrane structures," ed: Google Patents, 1996.

[28]   A. G. Mikos, G. Sarakinos, S. M. Leite, J. P. Vacant, and R. Langer, "Laminated three-dimensional biodegradable foams for use in tissue engineering," *Biomaterials,* vol. 14, pp. 323-330, 1993.

[29]   D. Mooney, S. Park, P. Kaufmann, K. Sano, K. McNamara, J. Vacanti, *et al.*, "Biodegradable sponges for hepatocyte transplantation," *Journal of biomedical materials research,* vol. 29, pp. 959-965, 1995.

[30]   K. Whang, C. Thomas, K. Healy, and G. Nuber, "A novel method to fabricate bioabsorbable scaffolds," *Polymer,* vol. 36, pp. 837-842, 1995.

[31]   Y. Y. Hsu, J. D. Gresser, D. J. Trantolo, C. M. Lyons, P. R. Gangadharam, and D. L. Wise, "Effect of polymer foam morphology and density on kinetics of in vitro controlled release of isoniazid from compressed foam matrices," *Journal of biomedical materials research,* vol. 35, pp. 107-116, 1997.

[32]   Y. S. Nam and T. G. Park, "Biodegradable polymeric microcellular foams by modified thermally induced phase separation method," *Biomaterials,* vol. 20, pp. 1783-1790, 1999.

[33]   H. Lo, S. Kadiyala, S. Guggino, and K. Leong, "Poly (L-lactic acid) foams with cell seeding and controlled-release capacity," *Journal of biomedical materials research,* vol. 30, pp. 475-484, 1996.

[34]   F. J. Hua, G. E. Kim, J. D. Lee, Y. K. Son, and D. S. Lee, "Macroporous poly (L-lactide) scaffold 1. Preparation of a macroporous scaffold by liquid–liquid phase separation of a PLLA–dioxane–water system," *Journal of biomedical materials research,* vol. 63, pp. 161-167, 2002.

[35]   H. Lo, M. Ponticiello, and K. Leong, "Fabrication of controlled release biodegradable foams by phase separation," *Tissue engineering,* vol. 1, pp. 15-28, 1995.

[36]   D. J. Mooney, D. F. Baldwin, N. P. Suh, J. P. Vacanti, and R. Langer, "Novel approach to fabricate porous sponges of poly (D, L-lactic-co-glycolic acid) without the use of organic solvents," *Biomaterials,* vol. 17, pp. 1417-1422, 1996.

[37] L. D. Harris, B. S. Kim, and D. J. Mooney, "Open pore biodegradable matrices formed with gas foaming," 1998.

[38] H. Yokoi, T. Kinoshita, and S. Zhang, "Dynamic reassembly of peptide RADA16 nanofiber scaffold," *Proceedings of the National Academy of Sciences of the United States of America*, vol. 102, pp. 8414-8419, 2005.

[39] S. Zhang, T. C. Holmes, C. M. DiPersio, R. O. Hynes, X. Su, and A. Rich, "Self-complementary oligopeptide matrices support mammalian cell attachment," *Biomaterials*, vol. 16, pp. 1385-1393, 1995.

[40] J. D. Hartgerink, E. Beniash, and S. I. Stupp, "Peptide-amphiphile nanofibers: a versatile scaffold for the preparation of self-assembling materials," *Proceedings of the National Academy of Sciences*, vol. 99, pp. 5133-5138, 2002.

[41] R. C. Thomson, M. J. Yaszemski, J. M. Powers, and A. G. Mikos, "Hydroxyapatite fiber reinforced poly (alpha-hydroxy ester) foams for bone regeneration," *Biomaterials*, vol. 19, pp. 1935-1943, 1998.

[42] S. J. Hollister, "Porous scaffold design for tissue engineering," *Nature materials*, vol. 4, pp. 518-524, 2005.

[43] M. E. Gomes, C. M. Bossano, C. M. Johnston, R. L. Reis, and A. G. Mikos, "In vitro localization of bone growth factors in constructs of biodegradable scaffolds seeded with marrow stromal cells and cultured in a flow perfusion bioreactor," *Tissue engineering*, vol. 12, pp. 177-188, 2006.

[44] M. E. Gomes, H. S. Azevedo, A. Moreira, V. Ellä, M. Kellomäki, and R. Reis, "Starch–poly (ε-caprolactone) and starch–poly (lactic acid) fibre-mesh scaffolds for bone tissue engineering applications: structure, mechanical properties and degradation behaviour," *Journal of Tissue Engineering and Regenerative Medicine*, vol. 2, pp. 243-252, 2008.

[45] S. H. Oh, S. G. Kang, and J. H. Lee, "Degradation behavior of hydrophilized PLGA scaffolds prepared by melt-molding particulate-leaching method: comparison with control hydrophobic one," *Journal of Materials Science: Materials in Medicine*, vol. 17, pp. 131-137, 2006.

[46] S. H. Oh, S. C. Park, H. K. Kim, Y. J. Koh, J.-H. Lee, M. C. Lee, *et al.*, "Degradation behavior of 3D porous polydioxanone-b-polycaprolactone scaffolds fabricated using the melt-molding

52   *Bioprinting*

particulate-leaching method," *Journal of Biomaterials Science, Polymer Edition,* vol. 22, pp. 225-237, 2011.

[47]   S. H. Oh, S. G. Kang, E. S. Kim, S. H. Cho, and J. H. Lee, "Fabrication and characterization of hydrophilic poly (lactic-< i> co</i>-glycolic acid)/poly (vinyl alcohol) blend cell scaffolds by melt-molding particulate-leaching method," *Biomaterials,* vol. 24, pp. 4011-4021, 2003.

[48]   F. Intranuovo, R. Gristina, F. Brun, S. Mohammadi, G. Ceccone, E. Sardella, *et al.,* "Plasma Modification of PCL Porous Scaffolds Fabricated by Solvent-Casting/Particulate-Leaching for Tissue Engineering," *Plasma Processes and Polymers,* 2014.

[49]   W. Lin, Q. Li, and T. Zhu, "Study of solvent casting/particulate leaching technique membranes in pervaporation for dehydration of caprolactam," *Journal of Industrial and Engineering Chemistry,* vol. 18, pp. 941-947, 2012.

[50]   D. Sin, X. Miao, G. Liu, F. Wei, G. Chadwick, C. Yan, *et al.,* "Polyurethane (PU) scaffolds prepared by solvent casting/particulate leaching (SCPL) combined with centrifugation," *Materials Science and Engineering: C,* vol. 30, pp. 78-85, 2010.

[51]   Q. Xing, X. Dong, R. Li, H. Yang, C. C. Han, and D. Wang, "Morphology and performance control of PLLA-based porous membranes by phase separation," *Polymer,* vol. 54, pp. 5965-5973, 2013.

[52]   C. V. Hoven, X. D. Dang, R. C. Coffin, J. Peet, T. Q. Nguyen, and G. C. Bazan, "Improved performance of polymer bulk heterojunction solar cells through the reduction of phase separation via solvent additives," *Advanced Materials,* vol. 22, pp. E63-E66, 2010.

[53]   A. Nandakumar, C. Cruz, A. Mentink, Z. Tahmasebi Birgani, L. Moroni, C. van Blitterswijk, *et al.,* "Monolithic and assembled polymer–ceramic composites for bone regeneration," *Acta biomaterialia,* vol. 9, pp. 5708-5717, 2013.

[54]   A. Gloria, R. De Santis, and L. Ambrosio, "Polymer-based composite scaffolds for tissue engineering," *Journal of Applied Biomaterials & Biomechanics,* vol. 8, 2010.

[55]   H. Ramanath, C. Chua, K. Leong, and K. Shah, "Melt flow behaviour of poly-ε-caprolactone in fused deposition modelling,"

Journal of Materials Science: Materials in Medicine, vol. 19, pp. 2541-2550, 2008.

[56] K. C. Ang, K. F. Leong, C. K. Chua, and M. Chandrasekaran, "Investigation of the mechanical properties and porosity relationships in fused deposition modelling-fabricated porous structures," *Rapid Prototyping Journal*, vol. 12, pp. 100-105, 2006.

[57] M. Too, K. Leong, C. Chua, Z. Du, S. Yang, C. Cheah, *et al.*, "Investigation of 3D non-random porous structures by fused deposition modelling," *The International Journal of Advanced Manufacturing Technology*, vol. 19, pp. 217-223, 2002.

[58] C. Lee, C. Chua, C. Cheah, L. Tan, and C. Feng, "Rapid investment casting: direct and indirect approaches via fused deposition modelling," *The International Journal of Advanced Manufacturing Technology*, vol. 23, pp. 93-101, 2004.

[59] S. J. Kalita, S. Bose, H. L. Hosick, and A. Bandyopadhyay, "Development of controlled porosity polymer-ceramic composite scaffolds via fused deposition modeling," *Materials Science and Engineering: C*, vol. 23, pp. 611-620, 2003.

[60] S. Ramakrishna, J. Mayer, E. Wintermantel, and K. W. Leong, "Biomedical applications of polymer-composite materials: a review," *Composites science and technology*, vol. 61, pp. 1189-1224, 2001.

[61] I. Zein, D. W. Hutmacher, K. C. Tan, and S. H. Teoh, "Fused deposition modeling of novel scaffold architectures for tissue engineering applications," *Biomaterials*, vol. 23, pp. 1169-1185, 2002.

[62] M. Endres, D. Hutmacher, A. Salgado, C. Kaps, J. Ringe, R. Reis, *et al.*, "Osteogenic induction of human bone marrow-derived mesenchymal progenitor cells in novel synthetic polymer-hydrogel matrices," *Tissue Engineering*, vol. 9, pp. 689-702, 2003.

[63] T. G. Van Tienen, R. G. Heijkants, P. Buma, J. H. de Groot, A. J. Pennings, and R. P. Veth, "Tissue ingrowth and degradation of two biodegradable porous polymers with different porosities and pore sizes," *Biomaterials*, vol. 23, pp. 1731-1738, 2002.

[64] F. Wang, L. Shor, A. Darling, S. Khalil, W. Sun, S. Güçeri, *et al.*, "Precision extruding deposition and characterization of cellular poly--caprolactone tissue scaffolds," *Rapid Prototyping Journal*, vol. 10, pp. 42-49, 2004.

[65]   Z. Xiong, Y. Yan, R. Zhang, and L. Sun, "Fabrication of porous poly(L-lactic acid) scaffolds for bone tissue engineering via precise extrusion," *Scripta Materialia,* vol. 45, pp. 773-779, 2001.

[66]   T. B. Woodfield, J. Malda, J. De Wijn, F. Peters, J. Riesle, and C. A. van Blitterswijk, "Design of porous scaffolds for cartilage tissue engineering using a three-dimensional fiber-deposition technique," *Biomaterials,* vol. 25, pp. 4149-4161, 2004.

[67]   Y. Yan, Z. Xiong, Y. Hu, S. Wang, R. Zhang, and C. Zhang, "Layered manufacturing of tissue engineering scaffolds via multi-nozzle deposition," *Materials Letters,* vol. 57, pp. 2623-2628, 2003.

[68]   P. D. Calvert and J. Cesarano III, "Freeforming objects with low-binder slurry," ed: US Patent 6,027,326, 2000.

[69]   G. Vozzi, A. Previti, D. De Rossi, and A. Ahluwalia, "Microsyringe-based deposition of two-dimensional and three-dimensional polymer scaffolds with a well-defined geometry for application to tissue engineering," *Tissue Engineering,* vol. 8, pp. 1089-1098, 2002.

[70]   Z. Xiong, Y. Yan, S. Wang, R. Zhang, and C. Zhang, "Fabrication of porous scaffolds for bone tissue engineering via low-temperature deposition," *Scripta Materialia,* vol. 46, pp. 771-776, 2002.

[71]   D. Therriault, S. R. White, and J. A. Lewis, "Chaotic mixing in three-dimensional microvascular networks fabricated by direct-write assembly," *Nature Materials,* vol. 2, pp. 265-271, 2003.

[72]   G. Vozzi, C. Flaim, A. Ahluwalia, and S. Bhatia, "Fabrication of PLGA scaffolds using soft lithography and microsyringe deposition," *Biomaterials,* vol. 24, pp. 2533-2540, 2003.

[73]   R. Flemming, C. Murphy, G. Abrams, S. Goodman, and P. Nealey, "Effects of synthetic micro-and nano-structured surfaces on cell behavior," *Biomaterials,* vol. 20, pp. 573-588, 1999.

[74]   J. W. Calvert and L. E. Weiss, "Assembled scaffolds for three dimensional cell culturing and tissue generation," 2000.

[75]   C. Chua, K. Leong, K. Tan, F. Wiria, and C. Cheah, "Development of tissue scaffolds using selective laser sintering of polyvinyl alcohol/hydroxyapatite biocomposite for craniofacial and joint defects," *Journal of Materials Science: Materials in Medicine,* vol. 15, pp. 1113-1121, 2004.

[76] K. Tan, C. Chua, K. Leong, C. Cheah, W. Gui, W. Tan, *et al.*, "Selective laser sintering of biocompatible polymers for applications in tissue engineering," *Bio-medical materials and engineering*, vol. 15, pp. 113-124, 2005.

[77] F. Wiria, K. Leong, C. Chua, and Y. Liu, "Poly-ε-caprolactone/hydroxyapatite for tissue engineering scaffold fabrication via selective laser sintering," *Acta Biomaterialia*, vol. 3, pp. 1-12, 2007.

[78] F. E. Wiria, K. F. Leong, and C. K. Chua, "Modeling of powder particle heat transfer process in selective laser sintering for fabricating tissue engineering scaffolds," *Rapid Prototyping Journal*, vol. 16, pp. 400-410, 2010.

[79] N. Sudarmadji, J. Tan, K. Leong, C. Chua, and Y. Loh, "Investigation of the mechanical properties and porosity relationships in selective laser-sintered polyhedral for functionally graded scaffolds," *Acta biomaterialia*, vol. 7, pp. 530-537, 2011.

[80] L. G. Griffith and G. Naughton, "Tissue engineering--current challenges and expanding opportunities," *Science*, vol. 295, pp. 1009-1014, 2002.

[81] K. Tan, C. Chua, K. Leong, C. Cheah, P. Cheang, M. Abu Bakar, *et al.*, "Scaffold development using selective laser sintering of polyetheretherketone–hydroxyapatite biocomposite blends," *Biomaterials*, vol. 24, pp. 3115-3123, 2003.

[82] K. Leong, F. Wiria, C. Chua, and S. Li, "Characterization of a poly-ε-caprolactone polymeric drug delivery device built by selective laser sintering," *Bio-medical materials and engineering*, vol. 17, pp. 147-157, 2007.

[83] T. Boland, A. Ovsianikov, B. N. Chickov, A. Doraiswamy, R. J. Narayan, W. Y. Yeong, *et al.*, "Rapid prototyping of artificial tissues and medical devices," *Advanced Materials & Processes*, vol. 165, pp. 51-53, 2007.

[84] R. L. Simpson, F. E. Wiria, A. A. Amis, C. K. Chua, K. F. Leong, U. N. Hansen, *et al.*, "Development of a 95/5 poly (L-lactide-co-glycolide)/hydroxylapatite and β-tricalcium phosphate scaffold as bone replacement material via selective laser sintering," *Journal of Biomedical Materials Research Part B: Applied Biomaterials*, vol. 84, pp. 17-25, 2008.

[85] F. E. Wiria, C. K. Chua, K. F. Leong, Z. Y. Quah, M. Chandrasekaran, and M. W. Lee, "Improved biocomposite development of poly (vinyl alcohol) and hydroxyapatite for tissue engineering scaffold fabrication using selective laser sintering," *Journal of Materials Science: Materials in Medicine,* vol. 19, pp. 989-996, 2008.

[86] F. E. Wiria, N. Sudarmadji, K. F. Leong, C. K. Chua, E. W. Chng, and C. C. Chan, "Selective laser sintering adaptation tools for cost effective fabrication of biomedical prototypes," *Rapid Prototyping Journal,* vol. 16, pp. 90-99, 2010.

[87] W. Yeong, N. Sudarmadji, H. Yu, C. Chua, K. Leong, S. Venkatraman, *et al.,* "Porous polycaprolactone scaffold for cardiac tissue engineering fabricated by selective laser sintering," *Acta Biomaterialia,* vol. 6, pp. 2028-2034, 2010.

[88] N. Vail, L. Swain, W. Fox, T. Aufdlemorte, G. Lee, and J. Barlow, "Materials for biomedical applications," *Materials & Design,* vol. 20, pp. 123-132, 1999.

[89] K. Leong, K. Phua, C. Chua, Z. Du, and K. Teo, "Fabrication of porous polymeric matrix drug delivery devices using the selective laser sintering technique," *Proceedings of the Institution of Mechanical Engineers, Part H: Journal of Engineering in Medicine,* vol. 215, pp. 191-192, 2001.

[90] K. Low, K. Leong, C. Chua, Z. Du, and C. Cheah, "Characterization of SLS parts for drug delivery devices," *Rapid Prototyping Journal,* vol. 7, pp. 262-268, 2001.

[91] S. S. Kim, H. Utsunomiya, J. A. Koski, B. M. Wu, M. J. Cima, J. Sohn, *et al.,* "Survival and function of hepatocytes on a novel three-dimensional synthetic biodegradable polymer scaffold with an intrinsic network of channels," *Annals of surgery,* vol. 228, p. 8, 1998.

[92] J. Zeltinger, J. K. Sherwood, D. A. Graham, R. Müeller, and L. G. Griffith, "Effect of pore size and void fraction on cellular adhesion, proliferation, and matrix deposition," *Tissue Engineering,* vol. 7, pp. 557-572, 2001.

[93] E. Sachlos and J. Czernuszka, "Making tissue engineering scaffolds work. Review: the application of solid freeform fabrication technology to the production of tissue engineering scaffolds," *Eur Cell Mater,* vol. 5, pp. 39-40, 2003.

[94] S. Bose, J. Darsell, M. Kintner, H. Hosick, and A. Bandyopadhyay, "Pore size and pore volume effects on alumina

and TCP ceramic scaffolds," *Materials Science and Engineering: C*, vol. 23, pp. 479-486, 2003.

[95]   T.-M. Chu, J. Halloran, S. J. Hollister, and S. E. Feinberg, "Hydroxyapatite implants with designed internal architecture," *Journal of Materials Science: Materials in Medicine*, vol. 12, pp. 471-478, 2001.

[96]   C. Chua, C. Feng, C. Lee, and G. Ang, "Rapid investment casting: direct and indirect approaches via model maker II," *The International Journal of Advanced Manufacturing Technology*, vol. 25, pp. 26-32, 2005.

[97]   W. Y. Yeong, C. K. Chua, K. F. Leong, M. Chandrasekaran, and M. W. Lee, "Development of scaffolds for tissue engineering using a 3D inkjet model maker," in *Virtual Modelling and Rapid Manufacturing: Advanced Research in Virtual and Rapid Prototyping Proc. 2nd Int. Conf. on Advanced Research in Virtual and Rapid Prototyping, 28 Sep-1 Oct 2005, Leiria, Portugal*, 2005, p. 115.

[98]   J. Taboas, R. Maddox, P. Krebsbach, and S. Hollister, "Indirect solid free form fabrication of local and global porous, biomimetic and composite 3D polymer-ceramic scaffolds," *Biomaterials*, vol. 24, pp. 181-194, 2003.

[99]   S. Limpanuphap and B. Derby, "Manufacture of biomaterials by a novel printing process," *Journal of Materials Science: Materials in Medicine*, vol. 13, pp. 1163-1166, 2002.

[100] C. Wilson, J. De Bruijn, C. Van Blitterswijk, A. Verbout, and W. Dhert, "Design and fabrication of standardized hydroxyapatite scaffolds with a defined macro-architecture by rapid prototyping for bone-tissue-engineering research," *Journal of Biomedical Materials Research Part A*, vol. 68, pp. 123-132, 2004.

[101] E. Sachlos, N. Reis, C. Ainsley, B. Derby, and J. Czernuszka, "Novel collagen scaffolds with predefined internal morphology made by solid freeform fabrication," *Biomaterials*, vol. 24, pp. 1487-1497, 2003.

[102] T.-M. G. Chu, D. G. Orton, S. J. Hollister, S. E. Feinberg, and J. W. Halloran, "Mechanical and in vivo performance of hydroxyapatite implants with controlled architectures," *Biomaterials*, vol. 23, pp. 1283-1293, 2002.

[103] B. H. Cumpston, S. P. Ananthavel, S. Barlow, D. L. Dyer, J. E. Ehrlich, L. L. Erskine, *et al.*, "Two-photon polymerization

initiators for three-dimensional optical data storage and microfabrication," *Nature,* vol. 398, pp. 51-54, 1999.

[104]   S. Maruo, O. Nakamura, and S. Kawata, "Three-dimensional microfabrication with two-photon-absorbed photopolymerization," *Optics letters,* vol. 22, pp. 132-134, 1997.

[105]   A. Ovsianikov, M. Malinauskas, S. Schlie, B. Chichkov, S. Gittard, R. Narayan, *et al.,* "Three-dimensional laser micro-and nano-structuring of acrylated poly (ethylene glycol) materials and evaluation of their cytoxicity for tissue engineering applications," *Acta biomaterialia,* vol. 7, pp. 967-974, 2011.

[106]   J. Serbin, A. Ovsianikov, and B. Chichkov, "Fabrication of woodpile structures by two-photon polymerization and investigation of their optical properties," *Optics Express,* vol. 12, pp. 5221-5228, 2004.

[107]   A. Ovsianikov, S. Schlie, A. Ngezahayo, A. Haverich, and B. N. Chichkov, "Two-photon polymerization technique for microfabrication of CAD-designed 3D scaffolds from commercially available photosensitive materials," *Journal of tissue engineering and regenerative medicine,* vol. 1, pp. 443-449, 2007.

[108]   S. H. Lee, J. J. Moon, and J. L. West, "Three-dimensional micropatterning of bioactive hydrogels via two-photon laser scanning photolithography for guided 3D cell migration," *Biomaterials,* vol. 29, pp. 2962-2968, 2008.

[109]   A. Ovsianikov, A. Ostendorf, and B. Chichkov, "Three-dimensional photofabrication with femtosecond lasers for applications in photonics and biomedicine," *Applied surface science,* vol. 253, pp. 6599-6602, 2007.

[110]   M. Centola, A. Rainer, C. Spadaccio, S. De Porcellinis, J. Genovese, and M. Trombetta, "Combining electrospinning and fused deposition modeling for the fabrication of a hybrid vascular graft," *Biofabrication,* vol. 2, p. 014102, 2010.

[111]   D. Espalin, K. Arcaute, D. Rodriguez, F. Medina, M. Posner, and R. Wicker, "Fused deposition modeling of patient-specific polymethylmethacrylate implants," *Rapid Prototyping Journal,* vol. 16, pp. 164-173, 2010.

[112]   E. Y. Teo, S. Y. Ong, M. S. Khoon Chong, Z. Zhang, J. Lu, S. Moochhala, *et al.,* "Polycaprolactone-based fused deposition

modeled mesh for delivery of antibacterial agents to infected wounds," *Biomaterials,* vol. 32, pp. 279-287, 2011.

[113]   C. Vaquette and J. Cooper White, "A simple method for fabricating 3-D multilayered composite scaffolds," *Acta biomaterialia,* vol. 9, pp. 4599-4608, 2013.

[114]   L. Shor, S. Güçeri, R. Chang, J. Gordon, Q. Kang, L. Hartsock, *et al.*, "Precision extruding deposition (PED) fabrication of polycaprolactone (PCL) scaffolds for bone tissue engineering," *Biofabrication,* vol. 1, p. 015003, 2009.

[115]   M. Domingos, D. Dinucci, S. Cometa, M. Alderighi, P. J. Bartolo, and F. Chiellini, "Polycaprolactone scaffolds fabricated via bioextrusion for tissue engineering applications," *International journal of biomaterials,* vol. 2009, 2009.

[116]   L. Liu, Z. Xiong, Y. Yan, R. Zhang, X. Wang, and L. Jin, "Multinozzle low-temperature deposition system for construction of gradient tissue engineering scaffolds," *Journal of Biomedical Materials Research Part B: Applied Biomaterials,* vol. 88, pp. 254-263, 2009.

[117]   J. Y. Kim, J. J. Yoon, E. K. Park, D. S. Kim, S. Y. Kim, and D. W. Cho, "Cell adhesion and proliferation evaluation of SFF-based biodegradable scaffolds fabricated using a multi-head deposition system," *Biofabrication,* vol. 1, p. 015002, 2009.

[118]   J. H. Shim, J. Y. Kim, J. K. Park, S. K. Hahn, J. W. Rhie, S. W. Kang, *et al.*, "Effect of thermal degradation of SFF-based PLGA scaffolds fabricated using a multi-head deposition system followed by change of cell growth rate," *Journal of Biomaterials Science, Polymer Edition,* vol. 21, pp. 1069-1080, 2010.

[119]   A. Tirella, F. Vozzi, G. Vozzi, and A. Ahluwalia, "PAM2 (piston assisted microsyringe): a new rapid prototyping technique for biofabrication of cell incorporated scaffolds," *Tissue Engineering Part C: Methods,* vol. 17, pp. 229-237, 2010.

[120]   G. Tartarisco, G. Gallone, F. Carpi, and G. Vozzi, "Polyurethane unimorph bender microfabricated with Pressure Assisted Microsyringe (PAM) for biomedical applications," *Materials Science and Engineering: C,* vol. 29, pp. 1835-1841, 2009.

[121]   G. Vozzi, A. Tirella, and A. Ahluwalia, "Rapid prototyping composite and complex scaffolds with PAM2," in *Computer-Aided Tissue Engineering,* ed: Springer, 2012, pp. 57-69.

[122]   B. Dorj, J. H. Park, and H. W. Kim, "Robocasting chitosan/nanobioactive glass dual-pore structured scaffolds for bone engineering," *Materials Letters,* vol. 73, pp. 119-122, 2012.

[123]   B. Dorj, J. E. Won, J. H. Kim, S. J. Choi, U. S. Shin, and H. W. Kim, "Robocasting nanocomposite scaffolds of poly (caprolactone)/hydroxyapatite incorporating modified carbon nanotubes for hard tissue reconstruction," *Journal of Biomedical Materials Research Part A,* vol. 101, pp. 1670-1681, 2013.

[124]   S. A. Park, S. H. Lee, and W. D. Kim, "Fabrication of porous polycaprolactone/hydroxyapatite (PCL/HA) blend scaffolds using a 3D plotting system for bone tissue engineering," *Bioprocess and biosystems engineering,* vol. 34, pp. 505-513, 2011.

[125]   J. M. Sobral, S. G. Caridade, R. A. Sousa, J. F. Mano, and R. L. Reis, "Three-dimensional plotted scaffolds with controlled pore size gradients: effect of scaffold geometry on mechanical performance and cell seeding efficiency," *Acta Biomaterialia,* vol. 7, pp. 1009-1018, 2011.

[126]   S. J. Hong, I. Jeong, K. T. Noh, H. S. Yu, G. S. Lee, and H. W. Kim, "Robotic dispensing of composite scaffolds and in vitro responses of bone marrow stromal cells," *Journal of Materials Science: Materials in Medicine,* vol. 20, pp. 1955-1962, 2009.

[127]   C. X. F. Lam, X. Mo, S.-H. Teoh, and D. Hutmacher, "Scaffold development using 3D printing with a starch-based polymer," *Materials Science and Engineering: C,* vol. 20, pp. 49-56, 2002.

[128]   B. Sharaf, C. B. Faris, H. Abukawa, S. M. Susarla, J. P. Vacanti, L. B. Kaban, *et al.,* "Three-dimensionally printed polycaprolactone and β-tricalcium phosphate scaffolds for bone tissue engineering: an in vitro study," *Journal of Oral and Maxillofacial Surgery,* vol. 70, pp. 647-656, 2012.

[129]   S. Eshraghi and S. Das, "Mechanical and microstructural properties of polycaprolactone scaffolds with one-dimensional, two-dimensional, and three-dimensional orthogonally oriented porous architectures produced by selective laser sintering," *Acta Biomaterialia,* vol. 6, pp. 2467-2476, 2010.

[130]   B. Duan, W. L. Cheung, and M. Wang, "Optimized fabrication of Ca–P/PHBV nanocomposite scaffolds via selective laser sintering for bone tissue engineering," *Biofabrication,* vol. 3, p. 015001, 2011.

[131] S. J. Kim, D. H. Jang, W. H. Park, and B.-M. Min, "Fabrication and characterization of 3-dimensional PLGA nanofiber/microfiber composite scaffolds," *Polymer*, vol. 51, pp. 1320-1327, 2010.

[132] F. Xu, S. Moon, A. Emre, E. Turali, Y. Song, S. Hacking, *et al.*, "A droplet-based building block approach for bladder smooth muscle cell (SMC) proliferation," *Biofabrication*, vol. 2, p. 014105, 2010.

[133] A. Nandakumar, H. Fernandes, J. de Boer, L. Moroni, P. Habibovic, and C. A. van Blitterswijk, "Fabrication of bioactive composite scaffolds by electrospinning for bone regeneration," *Macromolecular bioscience*, vol. 10, pp. 1365-1373, 2010.

[134] J. Schakenraad and P. Dijkstra, "Biocompatibility of poly(DL-lactic acid/glycine) copolymers," *Clinical materials*, vol. 7, pp. 253-269, 1991.

[135] A. Matsuo, T. Shuto, G. Hirata, H. Satoh, Y. Matsumoto, H. Zhao, *et al.*, "Antiinflammatory and chondroprotective effects of the aminobisphosphonate incadronate (YM175) in adjuvant induced arthritis," *The Journal of rheumatology*, vol. 30, pp. 1280-1290, 2003.

[136] M. H. Huo, D. W. Romness, and S. M. Huo, "Metallosis mimicking infection in a cemented total knee replacement," *Orthopedics*, vol. 20, p. 466, 1997.

[137] R. M. Donlan, "Biofilms and device-associated infections," *Emerging infectious diseases*, vol. 7, p. 277, 2001.

[138] L. Rimondini, M. Fini, and R. Giardino, "The microbial infection of biomaterials: a challenge for clinicians and researchers. A short review," *Journal of Applied Biomaterials & Biomechanics*, vol. 3, pp. 1-10, 2005.

[139] S. Fukagawa, S. Matsuda, H. Miura, K. Okazaki, Y. Tashiro, and Y. Iwamoto, "High-dose antibiotic infusion for infected knee prosthesis without implant removal," *Journal of Orthopaedic Science*, vol. 15, pp. 470-476, 2010.

[140] B. Pittet, D. Montandon, and D. Pittet, "Infection in breast implants," *The Lancet infectious diseases*, vol. 5, pp. 94-106, 2005.

[141] J. Costerton, P. S. Stewart, and E. Greenberg, "Bacterial biofilms: a common cause of persistent infections," *Science*, vol. 284, pp. 1318-1322, 1999.

[142] Y. Shinto, A. Uchida, F. Korkusuz, N. Araki, and K. Ono, "Calcium hydroxyapatite ceramic used as a delivery system for antibiotics," *Journal of Bone & Joint Surgery, British Volume*, vol. 74, pp. 600-604, 1992.

[143] J. Fu, J. Ji, W. Yuan, and J. Shen, "Construction of anti-adhesive and antibacterial multilayer films via layer-by-layer assembly of heparin and chitosan," *Biomaterials*, vol. 26, pp. 6684-6692, 2005.

Problems

1. What are the key requirements for fabrication of tissue engineering scaffolds?
2. What are the advantages and disadvantages of electrospinning?
3. Discuss the merits and drawbacks of Fused Deposition Modelling for scaffold fabrication.
4. What is indirect fabrication of tissue engineering scaffolds?
5. List all the laser-based additive manufacturing techniques for scaffold fabrication.

Chapter 3

Bioprinting Techniques

Three-dimensional (3D) cell bioprinting is an emerging technology, which has now been initially used to design and generate 3D cell constructs for transplantation therapies. The most attractive advantage of this technology is the ability to create 3D structures with living biological elements such as cells and nutrients. This chapter presents a number of companies with particular attention on their bioprinting systems and processes for delivering cells and biomaterials in preclinical studies, which have shown promising improvements as compared to the scaffold-based fabrication methods presented in Chapter 2. These systems are classified based on the printing techniques such as inkjet printing, extrusion and combinational techniques. Starting with the introduction of the bioprinting definition, this chapter describes the printing process principles and mechanisms, followed by discussion of the process advantages and disadvantages. Finally, the applications of the bioprinting techniques are outlined.

3.1 Bioprinting

3.1.1 *Definition*

Although there is no specific consensus on the definition of the term 'bioprinting', researchers have explored a number of approaches to deposit living cells and biomaterials with the similar objectives of directly creating precise 3D constructs. The definition of bioprinting was first proposed in the first bioprinting international conference at the University of Manchester Institute of Science and Technology (now part of the University of Manchester, UK) in September 2004. Bioprinting is defined as [1]:

'the use of material transfer processes for patterning and assembling biologically relevant materials – molecules, cells, tissues, and biodegradable biomaterials – with a prescribed organisation to accomplish one or more biological functions.'

This is a broad and open definition where bioprinting is considered as a set of techniques rather than a single approach. As long as a technique that is able to transfer biological materials onto a substrate ending up with 3D structures, it can be considered as a bioprinting technique. The ultimate aim of bioprinting is to manufacture living functional tissues and organs to be transplanted into human bodies. Mironov *et al.* [2] proposed another term 'organ printing' and a possible definition is [2]:

'computer aided 3D tissue engineering of living organs based on the simultaneous deposition of cells and hydrogels with the principles of self-assembly.'

3.1.2 *Overview of 3D bioprinting processes*

As stated in section 3.1.1, there are a number of existing 3D bioprinting techniques including extrusion, inkjet printing, valve-based printing, light processing, laser-based printing and combinational techniques. These bioprinting techniques cover a wide range of biological applications with different requirements such as patterning length scale, printing speed, cost and biocompatibility. Based on the material delivery methods, these printing techniques can be generally classified as contact and non-contact techniques.

• Contact: the printing technique requires contact between the delivery apparatus and the receiving substrate, such as the extrusion method.

- Non-contact: the material is delivered (ejected) to the substrate located very close to the delivery mechanism (almost touching). Typical examples are: laser-based and inkjet printing methods.

3.2 Extrusion I

3.2.1 *Company: EnvisionTEC*

EnvisionTEC Inc., founded in 2002 in Marl, Germany, is a world leader in Additive Manufacturing (AM) equipment, specialising in optical, mechanical, and electrical engineering [3]. EnvisionTEC develops and produces cost effective 3D printers including hardware, software and materials. Their products are widely used in the areas of aerospace, automotive, dental, education electronics, medical and bio-fabrication [4]. EnvisionTEC has gained high reputation in the manufacture of reliable AM system in the world using its core-based technology of selective light modulation. The simplicity of the technology has made the system very popular in AM markets such as the hearing aid market where EnvisionTEC enjoys more than 60% of the world market. The company owns over 90 patents and patent applications pending worldwide, and continues to issue an average of one patent every three months. EnvisionTEC is headquartered in Brüsseler Straße 51, D-45968 Gladbeck Germany, and has a number of branches in North America, Europe and Asia.

3.2.2 *Product: 3D-Bioplotter® System*

The 3D-Bioplotter® System is a suitable AM tool for processing a great variety of biomaterials within the process of Computer Aided Tissue Engineering (CATE) from 3D Computer Aided Design (CAD) models and patient Computed Tomography (CT) data to the physical 3D scaffold with a designed and defined outer form and an open inner structure. The 3D-Bioplotter® has the capability to fabricate scaffolds using materials ranging from soft hydrogels to hard ceramics and metals. Up to 5 materials can be used in printing an object and tool changes are fully

automated. The 3D-Bioplotter® is specially designed for work in sterile environments in a laminar flow box, a requirement of bioprinting, for example when using alginate cell suspensions for scaffold construction. A 3D-Bioplotter® machine, as shown in Fig. 3.1, adopts the most versatile AM machine configuration, namely 3-axis positioning system, realising 0.001 mm (0.00004 in.) positioning accuracy [5]. There are 5 different material cartridges that can be used during the same build and the strand diameter is controlled in real time via high resolution camera feedback, ensuring high part accuracy. In addition, EnvisionTEC developed a complete new CAD-CAM software application with an intuitive graphical user interface. Together with the dedicated software, the printing process is monitored until it is completed. The details of the 3D-Bioplotter® system are summarised in
Table 3.1. A number of materials that can be processed by the system for fabricating scaffolds are provided in Table 3.2.

Fig. 3.1. A 3D-Bioplotter® machine (courtesy of EnvisionTEC GmbH).

Table 3.1. Specifications of the 3D-Bioplotter®
system (courtesy of EnvisionTEC GmbH.

Axis Resolution (XYZ):	0.001mm
Speed:	0.1 - 150mm/s
Build Volume:	150 x 150 x 140mm
Needle Sensor Resolution (Z):	0.001mm
Camera Resolution (XY):	0.009mm per pixel
Minimum Strand Diameter:	0.100mm (material dependent)
Footprint:	976 x 625 x 773mm
Weight (approx):	130 kg
Electrical Requirement:	100 -240V AC, max. 3000VA, F 50/60Hz

Table 3.2. Available materials for the 3D-
Bioplotter® system.

Application area	Available material
Bone regeneration	Hydroxypapatite, Titanium and Tricalcium Phosphate (TCP)
Drug release	Polycaprolactone (PCL), Polylactic-Co-Glycolic Acid (PLGA) and Polylactic Acid (PLLA)
Soft tissue biofabrication/Organ printing	Agar, Chitosan, Alginate, Gelatine, Fibrin and Collagen

3.2.1 *Process and principle*

The 3D-Bioplotter® 3D printing technique deposits materials using air or mechanical pressure [6]. The pressure is applied to the syringes, each of which contains a material ranging from a viscous paste to a liquid. The material is deposited in a strand form to the substrate while the syringe is moving horizontally, as depicted in Fig. 3.2 where the strands are parallel to each other. The distance between each strand depends on the porosity defined by the user in the design stage. Upon finishing one layer, the moving direction of the syringe is turned over 90° and proceeds to print the next layer. This process continues until a physical representation of the CAD model is fully created. Figure 3.3 shows a finished object with a fine mesh.

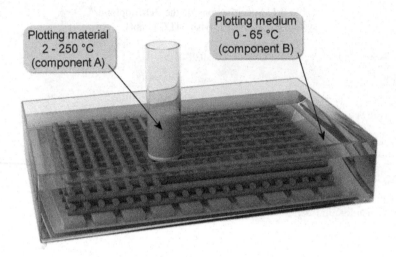

Fig. 3.2. The 3D-Bioplotter® printing process (courtesy of EnvisionTEC GmbH).

Fig. 3.3. A completed mesh (left) and the sectioned view (right) (courtesy of EnvisionTEC GmbH).

A few types of substrate/building platform are available, which are a cooled metal platform, a glass plate and a liquid. Obviously, printed cells or biomaterials can stack on solid plates. As for a liquid support, it has two special functions, namely, (1) enabling solidification through ionic transfer and other crosslinking approaches; (2) supporting deposited materials during the solidification process due to buoyancy resulting from the density difference between the material and the liquid.

Two important factors, i.e. strand thickness and interior structure design, need to be addressed in order to obtain high printing quality. The strand thickness directly determines the distance between the strand surface and the cell position, which plays an important role in cell proliferation [7]. In addition, the mechanical properties are largely influenced by the interior structure of the object. Therefore, certain changes should be made accordingly for depositing different materials.

3.3 Extrusion II

3.3.1 *Company: Organovo*

Organovo Holdings Inc., headquartered in San Diego USA, is a research-based incorporation. In early 2003, *Ink-jet Printing of Viable Cells* was patented by Dr. Thomas Boland at Clemson University. The Organ Printing work subsequently started at University of Missouri, developing an organ printing technology entitled *'Self-Assembling Cell-Aggregates and Methods of Making the Same'*, which provided the basis for the company formation. In April 2007, Organovo, Inc. formally incorporated in Delaware, with the intent to license the above patent suite and launch a 3D bioprinting company [8]. Organovo delivered their first commercial product named NovoGen MMX Bioprinter™ in September 2009. The company is now focusing on building a number of 3D tissue models for research and drug discovery applications, as well as working to fulfil the vision of building human tissues for surgical therapy and transplantation. The address of Organovo Inc. is 6275 Nancy Ridge Drive, Suite 110, San Diego, CA 92121.

3.3.2 *Product: NovoGen MMX Bioprinter™*

The Organovo's NovoGen MMX Bioprinter™ prints fully human, architecturally correct 3D tissue in a variety of different formats. This bioprinting platform, as shown in Fig. 3.4, is small and compact to fit easily into a standard biosafety cabinet, enabling sterile operation, and operates with minimal particle shedding. The platform consists of a precision robot and two deposition heads for dispensing different materials, enabling thick tissues (>500 microns in thickness) to be constructed with spatial control in the *X*, *Y*, and *Z* axes. Thus, tissue-specific patterns or compartments can be produced, which mimic key aspects of *in vivo* native tissues. The positional repeatability of the motion control system can achieve approximately ±10 μm.

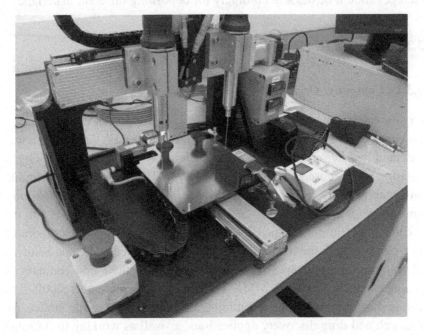

Fig. 3.4. The NovoGen MMX Bioprinter™ (courtesy of Organovo Inc.).

A built-in laser-based calibration unit guarantees high positioning accuracy between the two deposition heads in real time. This high precision and automation ensures reproducibility among bioprinted tissues through tight control of both the composition of the tissue and the geometry. The bioprinter is able to extrude spherical or cylindrical cellular aggregates with diameters of 500 or 260 μm, respectively, preloaded in the micropipette-cartridges (up to 75mm long) [9, 10], as shown in Fig. 3.5 and Fig. 3.6. In addition, hydrogels can also be printed as temporary and removable support structures for the cells during the printing process. Moreover, the platform is equipped with heating and cooling chambers capable of regulating temperatures from 4°C to 95°C. Thus, tissues can be fabricated directly into a wide variety of cultureware or custom chambers designed to maintain and condition 3D tissues. By doing so, the need for manipulations, which may introduce variability, can be significantly reduced. Figure 3.7 shows tissues being fabricated into a 24 well-plate by using a NovoGen MMX Bioprinter™ machine.

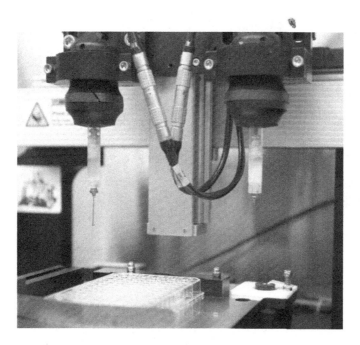

Fig. 3.5. Two dispensing heads for extruding bio-ink or hydrogels into multi-well plates (courtesy of Organovo Inc.).

Fig. 3.6. The bio-ink units packaged into capillary micropipette cartridges (left: spherical; right: cylinderical) [10] (copyright 2012, with permission of Springer).

Fig. 3.7. The NovoGen MMX Bioprinter™ fabricating tissue (courtesy of Organovo Inc.).

The software currently used only includes some basic functions. It has a graphical interface, which allows users to design various 3D constructs. The users can choose a number of parameters e.g. cell types and materials as well as printing speed. Furthermore, the software can load pre-defined commands written by the users for realising specific movements of the robot and deposition heads. Organovo is now collaborating with Autodesk (Autodesk Inc., San Rafael, CA, USA) to develop 3D bioprinting software to control their bioprinters.

3.3.3 *Process*

The first step is to develop the bioprocess protocols required to generate the multi-cellular building blocks, which is also called bio-ink, from the cells that will be used to build the target tissue. The printing process starts with extruding bio-ink units into the temporary support environment [11]. The bio-ink building blocks are dispensed from the bioprinter, using a layer-by-layer approach that is scaled for the target output. Bio-inert hydrogel components may be utilised as supports, as tissues are built up vertically to achieve three-dimensionality, or as fillers to create channels or void spaces within tissues to mimic features of native tissue. An example of bioprinting a blood vessel is shown in Fig. 3.8.

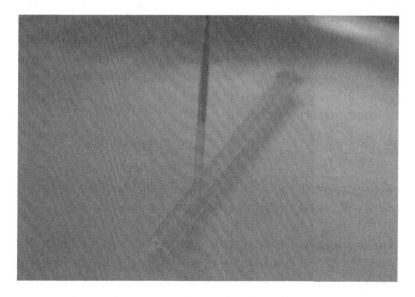

Fig. 3.8. The layer-by-layer bioprinting of a blood vessel, using hydrogel and cellular bio-ink in a precise spatial architecture (courtesy of Organovo Inc.).

Figure 3.9 illustrates the process of depositing spherical aggregates to create tubular structures. As the bioprinter houses multiple dispensing heads, the cellular bio-ink units and the agarose rods (support) can be generated by the same bioprinter. The bioprinting process can be further modified to produce tissues in a variety of formats, ranging from micro-scale tissues contained in standard multi-well tissue culture plates, to larger structures suitable for placement onto bioreactors for biomechanical conditioning prior to use.

Fig. 3.9. Design template for tubular structures (pink: agarose rods; orange: cylindrical bio-ink units) [12] (copyright 2009, with permission of Elsevier).

When the printing process is finished, the 3D construct will be subjected to post-processing. The post-processing stage takes place in an incubator, during which the structure is finally formed through fusing the printed bio-ink units. During the fusion process, the construct gradually becomes solid and then the support material is removed. Finally, the construct is moved to a bioreactor which provides near physiological conditions where maturation is promoted.

3.3.4 *Principle*

In order to explain the principle of this extrusion bioprinting process, the following cell behaviours/processes are first introduced.

- **Self-organisation** is defined as a process where patterning at the global level of a system emerges solely from numerous interactions among the lower-level components of the system [13]. Living organisms are examples of self-organising systems.

- *Self-assembly* is the autonomous organisation of components, from an initial state into the final patterns or structures without human intervention [14]. In other words, the morphological structure in embryonic development still evolves spontaneously in the self-assembly system without imposing an external impact to it. Self-assembly is a ubiquitous process in the field of developmental biology and organogenesis is a representative example.

- *Cell sorting* is a self-assembly process that provides a common mechanism to establish cellular compartments and boundaries between distinct tissues [15, 16]. The fluidic nature of cell sorting can be adequately explained by the theory of Differential Adhesion Hypothesis [17, 18]. If two different cell types are mixed randomly, the cells from the same types will gradually gather and aggregate whereas the distinct cell types separate due to different strength of adhesion between the neighbouring distinct cells.

- *Tissue fusion* is also a self-assembly process and in essence a phenomenon of fluid mechanics where two or more distinct types of cell make contact and coalesce [16, 19]. Therefore, the tissue fusion process is sometimes called 'melting together' and the motivation behind it is due to surface tension forces [17].

The Organovo's bioprinting process utilises the above cell behaviours and the principal concept of the bioprinting process is demonstrated in Fig. 3.10. The process of printing a tubular tissue construct is shown in Fig. 3.11 where the tissue spheroids are deposited layer-by-layer onto the cell-inert hydrogel supports, followed by tissue fusion and support removal.

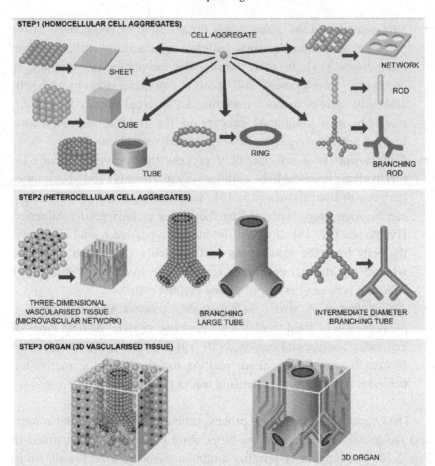

Fig. 3.10. The principal concept of the Organovo's extrusion printing technique [2] (copyright 2009, with permission of Elsevier).

Fig. 3.11. The printing process of a tubular tissue construct [2] (copyright 2009, with permission of Elsevier).

3.3.5 *Strengths and weaknesses*

The strengths of the extrusion-based Organovo's bioprinting process are:

(1) High efficiency. This extrusion bioprinting technique provides high printing speed, which is more efficient as compared to inkjet printing in the fabrication of large scale structures such as blood vessels [20].
(2) High cell density. Cell constructs with high level of cell density can be printed. Owing to the bioprinted tissues that are created without dependence on integrated scaffolding or hydrogel components, they have a tissue-like density with highly organised cellular features, such as intercellular tight junctions and microvascular networks.
(3) Feasible for thick tissue constructs. It can solve the problems of vascularisation in thick tissue constructs. Since the printed objects can be vascularised, thick tissue constructs can be bioprinted.
(4) Multiple compositions and geometries of 3D constructs. The Organovo's extrusion technique enables fabrication and comparative testing of multiple compositions and geometries so that winning combinations can be identified systematically based on histological and functional outcomes.
(5) Less dependency. In comparison to standard cell-culture platforms, Organovo's 3D bioprinted tissues are not dependent on biomaterial or scaffold components, which cannot be found in native tissues.

By contrast, the major drawbacks of this technique are as follows:

(1) Controlled environment is required. For fabricating a large construct, the printed cells may not survive before the lengthy printing process is complete. A controlled microenvironment which is able to maintain temperatures, provide oxygen and deliver nutrition etc. is required.
(2) Spheroids of standard size are required.
(3) Lack of stability in vertical dimension, and unable to print tall constructs.

(4) Being able to print long constructs by using support materials, but unable to readily remove support materials.

(5) Shrinkage of construct after tissue fusion.

(6) Spheroid fusion is required prior to printing.

3.4 Extrusion III

3.4.1 *Company: GeSiM*

GeSiM is a non-public and privately held company, founded in 1995 as a spin-off from the Rossendorf Research Centre. With a broad expertise in micromachining technologies as their solid support, GeSiM has evolved into a bioinstrumentation company specialising in microfluidics, sub-microliter liquid handling instruments and micro-contact printing. GeSiM offers complete tailored solutions to customers, including microfluidics, macrofluidcs, mechanics, packaging and specialised control software.

3.4.2 *Product: BioScaffolder*

The BioScaffolder, as shown in Fig. 3.12, is a modular instrument platform which can house up to four independent Z axes for dispensing materials. It is able to print scaffolds as well as living cells. The printable materials include biopolymers (e.g. collagen and alginate), hydrogels, bone cement paste, polymer pastes and biocompatible silicones. The working area is 100 × 346 × 40 mm. BioScaffolder provides piezoelectric nano-litre pipetting and pressure-driven 3D printing on the same instrument. The print head operates up to three pneumatic dispense cartridges for depositing highly viscous paste. Each dispenser is controlled by an individual Z-stepper motor with step width of 2 μm in X/Y axes and 10 μm in Z axis. The independent Z-drives are able to print different materials without changing cartridges. The nano-litre pipetting system is able to pipet up to 384 sample species on parts of a scaffold structure. The configuration with two cartridges and a piezoelectric dispenser is shown in Fig. 3.13. The key features of the BioScaffold system are summarised in Table 3.3 and the main applications include: printing of living cells (either seeded by a piezoelectric micro-dispenser or embedded in the scaffold material), producing soft tissue implants, plotting of conductive coatings onto medical devices, dispensing of photosensitive materials and manufacture of sensors from conductive polymers.

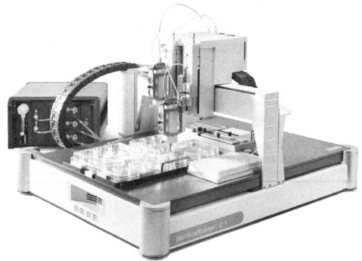

Fig. 3.12. A BioScaffolder platform (courtesy of GeSiM mbH).

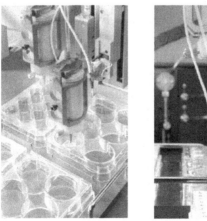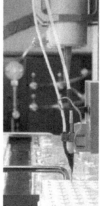

Fig. 3.13. The configuration with two cartridges and a piezo dispenser (courtesy of GeSiM mbH).

Table 3.3. Key features of the BioScaffold
system (courtesy of GeSiM mbH).

Feature	Function
Stage	Gantry system with accurate but smooth belt drives; spindle drives are used for Z axes
Cartridge	Up to three cartridges can be used in one single setup; each Z axis comes with a 30mm cartridge holder
Cartridge heater	Highly viscous materials can be deposited at elevated temperatures. The cartridge heater accommodates 10 mL cartridges; the maximum temperature is 180°C
FluidicBox	Independent pressure control is provided for each cartridge; quick and accurate start/stop cycles are realised by using additional valves in the print head
Software	Cubic and cylindrical 3D objects can be defined in GUI-based control software; each cartridge can be assigned to a plot strand. Module 1: scaffold generator; Module 2: STL-interface

3.5 Extrusion IV

3.5.1 *Company: Cyfuse Biomedical K.K.*

Cyfuse Biomedical K.K. is a start-up company headquartered in Tokyo, Japan. The company was founded by Koji Kuchiishi and Koichi Nakayama in August 2010. Cyfuse specialises in the development and manufacture of 3D tissue products. The company also endeavour to develop manufacturing systems for fabricating 3D tissue products. When Cyfuse was initiated, the bio 3D printer was developed in collaboration with Shibuya Kogyo Co., Ltd. A year later, Cyfuse and Kyushu University launched the research on osteochondral regeneration. In December 2012, the company formally introduced a brand new 3D printing system named Regenova®.

3.5.2 Product (Regenova®) and Process

Regenova® is an automatic robotic system that creates 3D cellular structures by depositing and placing cellular spheroids in needle arrays. The printing process is similar to that of Organovo's NovoGen MMX Bioprinter™, which can be divided into three major steps outlined in Table 3.4.

Table 3.4. The major steps for printing a 3D construct (courtesy of Cyfuse Biomedical K.K.).

Step		Process Description	Visual Demonstration
1	Loading cell spheroids and preparing the 3D data and needle array	Spheroids: cells are isolated and proliferated to required levels. Typically, a cellular spheroid is around 500 µm in diameter, consisting of tens of thousands of cells.	
		3D data: 3D arrangement of spheroids is designed in the Cyfuse's software. Various types of spheroid can be chosen in one design.	
		Needle array: needles made of stainless steel with 100-200 µm in diameter are used to act as a temporary support (i.e. scaffold) during spheroid assembly.	
2	3D printing	Spheroids are assembled into a 3D construct by depositing and placing them in the prepared needle arrays.	
3	Maturation	3D printed tissue construct is cultured in a bioreactor system to allow cell rearrangement and organisation, by which the tissue functions and physical strength are obtained.	

3.6 Inkjet Printing I

Inkjet printing is a non-contact printing technique that takes data from a computer representing an image or character, and reproduces it onto a substrate by ejecting tiny ink drops on a drop-on-demand manner [21]. Drop-on-demand indicates that the ink is discharged onto the substrate only where and when it is actually required to generate the constructs. Inkjet printing is not a new technology and it has been widely used in electronics and micro-engineering industries for printing electronic materials and complex integrated circuits [22]. Recently, Boland *et al.* [23] reported that the inkjet technology has gradually been adapted and applied to printing cells and biomaterials such as hydrogels in medicine and biomedical fields. In this chapter, the term 'inkjet printing' refers to inkjet bioprinting rather than the traditional inkjet printing methods.

3.6.1 *Company: Fujifilm*

Fujifilm Holdings Corporation Ltd., commonly known as Fujifilm (established in 1934), is a Japanese multinational photography and imaging company headquartered in Tokyo, Japan. The bioprinting system to be introduced in the next section is released by Fujifilm USA. Fujifilm companies in the USA serve a broad spectrum of industries including medical, graphic arts, optics, enterprise storage, motion picture and photography. For the past 78 years Fujifilm has continually invested in research and development resulting in world-class, highly versatile fundamental core technologies. The holding company for the USA-based Fujifilm companies operates in 27 states and engages in the research, development, manufacture, sales and services of Fujifilm products. The Fujifilm developed technologies spread in the areas of think film formation and processing, organic and inorganic materials, optics, imaging, drug development, mechatronics and electronics. The registered headquarter of Fujifilm is in 2-26-30 Nishiazabu, Minato-ku, Tokyo 106-0031, Japan.

3.6.2 *Product: Dimatix Materials Printer (DMP)*

The DMP-2831 allows the deposition of fluidic materials (e.g. biological fluids including cell patterning, DNA arrays, proteomics) on an 8 × 11 inch or A4 substrate, utilising a disposable piezoelectric inkjet cartridge. This printer can create and define patterns over an area of about 200 × 300 mm^2 and handle substrates up to 25 mm thick with an adjustable Z height [24]. The temperature of the vacuum platen, which secures the substrate in place, can be adjusted up to 60°C. The DMP-2831 offers a variety of patterns using a pattern editor program. Moreover, a waveform editor and a drop-watch camera system allow manipulation of the electronic pulses to the piezoelectric jetting device for optimisation of the drop characteristics as it is ejected from the nozzle. This system enables easy printing of structures and samples for process verification and prototype creation. The dedicated Graphical User Interface (GUI) application software which comes with the printer, imports Bitmap files as the CAD models. The software provides a conversion function to convert DXF, Gerber, GDSII and OASIS file conversion to Bitmap format. A DMP-2831 machine is shown in Fig. 3.14 and the specifications of the mechanical system is summarised in Table 3.5.

Fig. 3.14. A Fujifilm's Dimatix Materials Printer-2831 (courtesy of Fujifilm Ltd.).

Table 3.5. The specifications of the DMP-2831
mechanical system (courtesy of Fujifilm Ltd.).

Machine specification	Dimatix Materials Printer-2831 parameter
Printable area	Substrate < 0.5 mm thickness:
	210 mm × 315 mm (8.27 in × 12.4 in)
	Substrate 0.5 – 25 mm thickness:
	210 mm × 260 mm (8.27 in × 10.2 in)
Repeatability	± 25 µm (± 0.001 in)
Resolution	5 – 254 µm dot pitch (100 – 5080 dpi)
Substrate holder	Vacuum platen
	Temperature adjustable; ambient to 60°C
System footprint	673 mm × 584 mm× 419 mm (26 in × 23 in × 16 in)
Weight (approximately)	43 kg (95 lbs)
Power	100-120/200-240 VAC, 50/60 Hz, 375 W maximum
Operating range	15-40°C at 5-80% RH non-condensing

The most unique feature of the DMP-2831 printing system is the printhead, which allows users to fill their own fluids and print straightaway. To minimise waste of expensive fluids, each cartridge (Dimatix Materials Cartridge, DMC) reservoir has a capacity of 1.5 ml. Cartridges can easily be replaced to facilitate printing of a series of fluids. Each single-use cartridge has 16 nozzles linearly spaced at 254 microns with typical drop sizes of 1 and 10 picoliters. Cartridges are designed for high-resolution, non-contact jetting of functional fluids in a broad range of applications. A cartridge with drop size of 1 picoliters can deposit features as small as 20 µm, allowing users to closely pack large numbers of elements in DNA arrays and cell constructs. Figure 3.15 shows a replaceable print cartridge with one-time user fillable reservoir and its major features are outlined in Table 3.6.

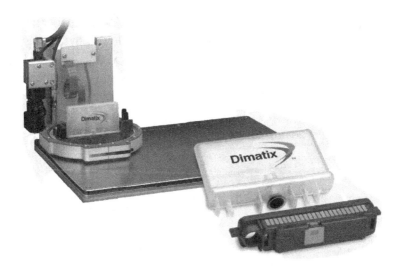

Fig. 3.15. A DMP-2831 print cartridge (courtesy of Fujifilm Ltd.)

Table 3.6. Specifications of the print cartridge for
the DMP-2831 system (courtesy of Fujifilm Ltd.).

Cartridge specification	Print cartridge parameter
Type	Piezo-driven jetting device with integrated reservoir and heater
Usable ink capacity	Up to 1.5 ml (user-fillable)
Material compatibility	Many water-based, solvent, acidic or basic fluids
Number of nozzles	16 nozzles, 254 μm spacing, single row
Drop volume	1 (DMC-11601) and 10 (DMC-11610) picoliter nominal

3.6.3 *Process and principle*

Inkjet bioprinting works in a similar manner to the inkjet technology applied in most of the commercial desktop printers. The thermal and piezoelectric are the most common types that inkjet systems adopt. The method for both types to print cells is very similar and is depicted in Fig. 3.16. A 3D cell construct is fabricated layer-by-layer from the bottom to the top. The 'structural' cell printing approach (Fig. 3.16 left) uses the same deposition tool to create scaffolding, cells and biomolecules in sequence (or simultaneously if multiple deposition tools are available).

Another printing method (Fig. 3.16 right), termed 'conformal', first lays down a thin layer as a prefabricating scaffold, followed by adding cells and biomolecules onto the top of the layer. The next thin layer will then be built to support the following cells. This fabrication cycle continues until the 3D structure is generated.

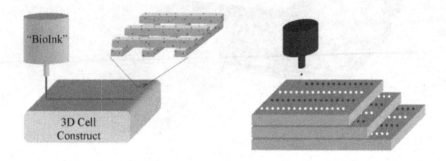

Fig. 3.16. Schematic of cell printing structural and conformal approaches [25] (Copyright 2006, with permission of John Wiley and Sons).

For a piezoelectric inkjet printing process, ink drops are ejected through actuation generated by the piezoelectric actuator in the reservoir. A short current pulse is applied to a piezoelectric element instead of the heater, leading to a shape change of the fluid reservoir. The fluid is ejected from the nozzle as a result of the reservoir contraction. After jetting the fluid, the reservoir regains its original shape and is refilled for the next ejection.

In a thermal inkjet system, the basic elements are a heating unit and an ink chamber or reservoir with a number of small nozzles (orifice diameter of 30–200µm) [26]. A controlled short pulse is first applied to the heater to significantly raise the heater temperature as high as 300°C within 10 microseconds on the surface. This results in the increased bulk temperature of the fluid by 5-10°C. As a result, a small air bubble is created, which subsequently expands and collapses. The bubble collapse provides the pressure pulse, forcing a tiny drop of ink to eject out of the nozzle [27]. The chamber is refilled with ink for the next ink ejection.

In general, the overall inkjet printing process can be described as two stages [28], although some researchers consider it as a single stage (two stages take place simultaneously) [29]. The first stage is to plate a thin layer made of biomaterials onto the substrate, functioning as a scaffold to support the cells to be printed in the second stage. This procedure is repeated until the 3D cellular structure is constructed. In detail, an inkjet printing process can be divided into four continuous steps, which are:

(i) generating pressure and getting ready to eject fluids through heating the heater (thermal inkjet) or piezoelectric actuation;

(ii) a bubble is formed and subsequently collapsed, leading to a droplet to be ejected through the orifice;

(iii) the formed droplet is deposited onto a substrate;

(iv) the ink-jet mechanism is automatically recovered to its original configuration for the next droplet deposition.

3.6.4 *Strengths and Weaknesses*

The major advantages of inkjet printing are outlined as follows:

(1) Low operation cost, high reproducibility and non-contact deposition [20]. Inkjet printing is able to create multiple identical 3D cell constructs. It will also be beneficial from using inkjet printing in the experiments where the material consumption is low i.e. low reagent costs.

(2) High automation. Computer controlled movements of printheads enable bio-ink and cells to be accurately deposited. The deposition of bio-ink layers is critical because each layer serves as the support/surface for generating the 3D backbone [30].

(3) Organ printing capability. The ability to place a wide variety of hydrogel scaffolds comprising of natural or synthetic polymers facilitates the direct printing of organs and tissues.

The challenges that inkjet printing faces are summarised below:

(1) Nozzle orifice clogging. Viscous cells can result in the clogging risk of the nozzle orifice, which in turn, increases the shear of the ejected cells at the nozzle. This high mechanical shear can lead to the damage of the cells.

(2) Cell degradation. During droplet formation and inkjet mechanism recovery, hydrostatic and inertial forces develop, which leads to various forms of cell degradation.

(3) Difficulties in cell aggregation and sedimentation in the printer reservoir and tubing. Due to cells that are required to be suspended in the cartridge reservoir before extrusion, cell aggregation and sedimentation become an inherent drawback of the inkjet printing technology [31], which cannot be completely eliminated.

(4) High shear strain during jetting. It is obvious that the fluid flow at the nozzle orifice centre is significantly higher than the flow in other places in the tube such as orifice walls. This velocity gradient leads to high shear flow at the orifice where the fluids are ejected, causing rupture of the living cells.

(5) High 'impact' with substrates. The interaction between the ejected droplet and the receiving substrate is considered to be an 'impact' [32]. High droplet jetting velocities can induce forces between the droplets and the substrate, which is detrimental to cells.

(6) Relatively low resolution. The resolution of inkjet printing is 100 µm, which is largely limited by the size of the nozzle orifice. In addition, the ejected droplet diameter is nearly 200% larger than that of the orifice. However, using fine nozzles is most likely to induce orifice blockage.

(7) Not being able to print a high-cell density constructs.

3.7 Inkjet Printing II

3.7.1 *Company: Microjet Corporation*

Microjet Corporation is a Japanese company in the field of inkjet technology. The company was founded in September 1997 and their headquarters is now in Nagagno-ken, Japan. Microjet endeavours to explore advanced, innovative and unique technology development, create new values by organically connecting those results with other technologies from customers and continuously develop new environmentally friendly products and launch to global market. The major services Microjet provides include device testing, development of inkjet printer for industrial application and droplet observation and measuring device.

3.7.2 *Product: LabJet-Bio System*

LabJet-Bio is a piezo inkjet high precision dispensing device. It sucks a tiny amount of liquid and dispenses or patterns on demand. It is also able to handle various kinds of liquid, with vessels equipped up to 16 units. A LabJet-Bio system is shown in Fig. 3.17 and its major features are as follows:

- Automatically sucking tiny amount of liquid, and dispensing and patterning (minimum required volume : 0.2ml)
- Multiple liquid handling controlled by automatic rinse system
- Liquid droplets monitoring function (see Fig. 3.18)
- Work alignment function, table observation function
- Piezo head that can handle high surface tension or high viscosity liquid

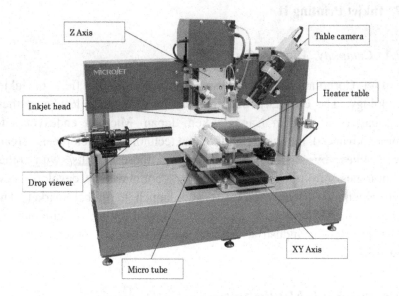

Fig. 3.17. A LabJet-Bio system (courtesy of Microjet Corporation).

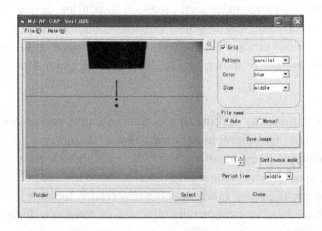

Fig. 3.18. The monitoring function of ejecting droplet (courtesy of Microjet Corporation).

The LabJet-Bio system consists of a main unit providing XYZ axes automatic control, a stage controller, a head controller, an embedded head and a liquid sucking single nozzle piezo head. Table 3.7 outlines the LabJet-Bio system specification. The major application areas include dispensing and patterning of protein, anti-body, enzyme, cells and reagents, producing bio-chips and bio-sensors, circuit design with nano-metal ink, drug screening and manufacturing testing of cell sheets.

Table 3.7. The specification of a LabJet-Bio system (courtesy of Microjet Corporation).

Parameter	Specification
Machine size	Approx. W470×D340×H370
Patterning area	80×40 mm *varies by model
Patterning function	dots, lines, face (by continuous line drawing), dispense fold function
Patterning accuracy	XY returning position accuracy ±5 μm
Droplet observation function (standard)	flushing parts, CCD, flush control embodied in head controller
Eject liquid volume	10 – 3,000 pl/droplet * depend on heads and liquids

3.7.3 *Process and Principle*

The process flow of using the bioprinting system is presented below:

- Set cell liquid to micro tube
- Minimum cell liquid volume to set : 0.2cc
- Set the positioning of vessel containing cell liquid and sack them by certain amount automatically
- Check the status of ejecting droplet, by PC display
- Select the position on the table to eject droplet using the software
- Select the number of droplets containing cells to eject
- Execute ejection
- Apply pressure to cell liquid remaining in head which will then return to designated vessel

Two modes are specifically developed by Microjet, which are 'Pull-Push' mode that is commonly used for consumer inkjet printer and 'Push-Pull' mode that fits to cell ejection. Users can select the proper mode according to different printing requirements.

3.7.3.1 *Pull-Push method*

'Pull-Push' method is used for all generic inkjet printers in order to minimise droplet size, or driving voltage for the piezo element. However, it is not the case to consistently eject cells. 'Pull-Push' mode generates large deformation of meniscus and thus sucking bubbles. It becomes unstable on ejecting liquid containing particles such as cells.

Figure 3.19a shows 'Pull-Push' voltage waveform to be applied to the piezo element and Fig. 3.19b shows the actual driving operation of the piezo element. For 'Pull-Push' mode, voltage (Vh) is applied to the piezo element on stand-by mode (①). In this status, the piezo element is bent to pressure chamber side, but no pressure happens, since it remains stationary. By giving an ejection signal, the electrical charge stored in the piezo element is released, and the voltage applied to the piezo element becomes 0 (②). This operation is defined as Pull mode. The piezo element returns to its original shape by an electric discharge. In this process, a negative pressure occurs in the pressure chamber, because the volume of the pressure chamber grows. The meniscus in the nozzle generated by the negative pressure is pulled strongly towards the inside of the nozzle. Then the negative pressure turns into the positive pressure, and is transmitted to the edge of nozzle. Right after the timing when the positive pressure wave resonates (Pw), the electric charge is applied again to the piezo element, making it bent and transformed (④). This operation is defined as Push mode. This operation generates a large pressure and ejects liquid droplets with high speeds through the nozzle.

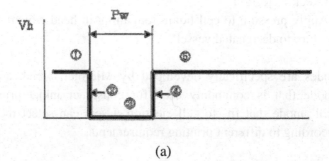

(a)

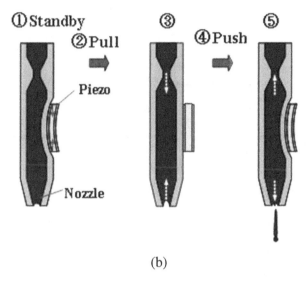

(b)

Fig. 3.19. Pull-Push method (courtesy of Microjet Corporation).

3.7.3.2 *Push-Pull method*

'Push-Pull' mode is a countermeasure and the voltage waveform is depicted in Fig. 3.20a. The actual operation process of the piezo element is illustrated in Fig. 3.20b.

For Push-Pull mode, voltage (Vh) is not applied to the piezo element on stand-by mode (①). By giving an ejection signal, the voltage is applied to the piezo element, and then the shape is transformed and becomes Push mode (②). Subsequently, a positive pressure occurs, by compressing the pressure chamber. This pressure wave enables the ejection of liquid droplets through the nozzle. The piezo element retains the original status by applying a pulse width (Pw) (③). The electric charge in the piezo element is released. The voltage becomes 0 and turns back to Pull mode (④) and displacement becomes 0 and returns to stand-by mode.

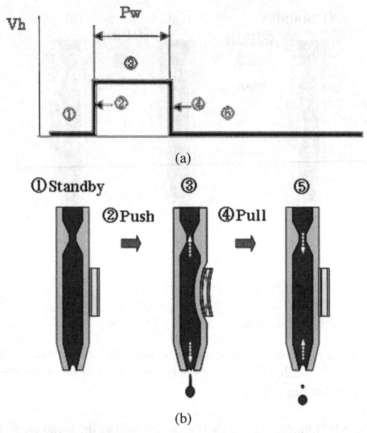

Fig. 3.20. Push-Pull method (courtesy of Microjet Corporation).

3.7.4 *Strengths and Weaknesses*

The advantages of the LabJet-Bio system can be summarised as follows.

- Stable ejection of cell liquid (clogging free)
- High density cell liquid applicable (1×10^6 cells/cc)
- Stably ejects cells up to 30 μm diameter
- Real-time monitoring of the status of cell ejection, by PC display
- Applicable to tiny amount of cell liquid
- Functions to disperse cells which tends to subside

The weaknesses are similar to section 3.6.4.

3.8 Light Processing

3.8.1 *Company: RegenHU*

RegenHU Ltd., located in Switzerland, is an innovative biomedical company for tissue engineering and therapy research. They joined the CPA Group (CPA Group SA, Switzerland) in 2011. RegenHU specialises on bio-system design and development with the primary aim of exploiting new bioprinting solutions in order to respond to the emerging challenges facing the biomedical industry. RegenHU benefits from exclusive patented technologies resulting from many years of research within international universities and partners. The company address is Z.i. du Vivier 22, 1690 Villaz-St-Pierre, Switzerland.

3.8.2 *Products: BioFactory® and 3DDiscovery®*

The BioFactory® is a high end, versatile and cell friendly 3D bioprinting instrument. It allows scientists to pattern cells, biomolecules and a range of soft and rigid materials in desirable 3D composite structures in order to mimic biomimetic tissue models [33]. The BioFactory® instrument provides a powerful tool for tissue engineering to create organotypic tissues with *in vivo*-like morphology that better mirror the environment experienced by cells *in vivo*. Furthermore, it better reflects cell behaviour, intercellular interactions and differentiation processes.

As an entry level product, the 3DDiscovery® instrument is a cost effective 3D bioprinting platform to explore the potential of 3D tissue engineering through the bioprinting approach. Spatial control of cells and morphogens in a 3D scaffold is an innovative approach to construct designed organotypic *in vitro* models of soft and hard tissues. The prototypes of BioFactory® and 3DDiscovery® are shown in Fig. 3.21 and the specification overview of 3DDiscovery® can be found in Table 3.8.

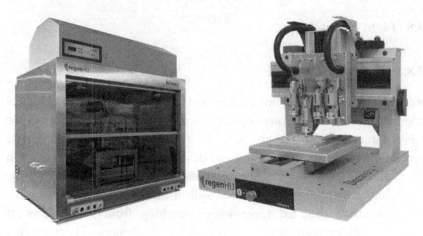

Fig. 3.21. The BioFactory® (left) and 3DDiscovery® (right) (courtesy of RegenHU Ltd.).

Table 3.8. The specifications of the 3DDiscovery®.

Specification	3DDiscovery® model
External dimensions	580/540/570 mm
Precision	±10 μm
Working range	130 mm × 90 mm × 60 mm
Viscosity range	Up to 10'000 mPaS
Temperature control	Up to 80°C
Available materials	Biopolymers, calcium, cells, signal molecules (proteins), hydrogels, collagen, polycaprolactone, polyester
Number of dispenser channels	4
Dispensing spectrum	10 nl up to ml

The Windows-based Human Machine Interface (HMI) software is used to control BioFactory® systems and the main window is shown in Fig. 3.22. The main window is divided into four parts, which are operation tabs, hardware configuration, manipulator control, and status and error bar. Operation tabs show the information relative to the selected operation tab such as Auto, Manual, Settings and Tools. Hardware configuration informs the user about the hardware configuration and its parameterisation. Manipulator control is used to switch on/off the power of the manipulator and to unlock and open the drawer with the building platform. Status and Error bar displays process status and error messages.

Fig. 3.22. The main window of the Human Machine Interface (courtesy of RegenHU Ltd.).

3.8.3 *Major feature: combination*

BioFactory® is designed for research purposes, and thus it combines both contact and non-contact printheads. The printheads are classified based on material viscosity, e.g. inkjet for low viscosity, extrusion for medium to high viscosity, and melt extruder for solid. Ultraviolet (UV) and laser beam are integrated in the system, enabling photo-sensitive polymers to be printed in a liquid form and cured to solid right after printing.

3.9 Valve-Based Printing I

3.9.1 *Company: Digilab*

Digilab Inc., headquartered in Marlborough, America, is a leading innovation company in spectrometry and photonics spanning over the past four decades. Digilab was formed and was a pioneer in Fourier Transform InfraRed Spectroscopy. In the mid-1970s Digilab was sold to Bio-Rad Laboratories which, in 2001, was purchased by Sidney

Braginsky who re-launched Digilab LLC. The company designs, develops and manufactures liquid handling tools, imaging and spectroscopy products for sample identification. Digilab Inc. is now focused on advanced systems and solutions in the high growth life sciences research markets, in particular, development of spectroscopy and alternatives. Upon experiencing a series of acquisitions and partnerships from 2004 to 2007, Digilab has further extended its global footprint including clinical diagnostics, analytical chemistry, safety and security applications. The company has moved to a new premise, 100 Locke Drive, Marlborough, MA, 01752, USA.

3.9.2 Product: CellJet Printer

The CellJet Cell Printing system integrates Digilab's proprietary synQUAD liquid dispensing technology [34], capable of printing cells in a non-contact manner. The flexible dispensing technology gives users full control over critical dispense parameters such as height of dispense and dispensing speed, which allows printing of both viscous solutions and fragile cells [35]. The CellJet cell printer designed to deposit cells onto slides or plates using a valve-free fluid path and thus, the risks of cell damage are largely reduced. CellJet is equipped with a sterilised evaporation cover, which is very easy to clean. A CellJet printer is shown in Fig. 3.23 and the key features are outlined as follows:

• The system can be specifically configured according to different application requirements. For example, one-channel configuration for printing cells from single cell type; two-channel configuration for printing multiple cell types or both cell and reagent; and four-channel configuration for bulk dispensing of cells.
• Multiple channels are available for multiple cell lines and multiplex assays are cross contamination free.
• The system together with synQUAD technology is able to dispense full range of cell types, from robust to delicate, and fluid types from low viscosity solvent to high viscous solutions such as 30% PEG.
• The dispensing speed and dispense height can be defined by operator to improve printing quality for different types of cell.

- Two dispensing modes are provided, namely, drop-by-drop and on-the-fly (to be introduced in the next subsection) for different throughput requirements.

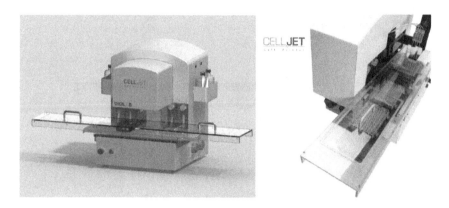

Fig. 3.23. A CellJet Cell Printer (courtesy of Digilab Inc.).

The specifications of the CellJet printer are given in Table 3.9. The specifically developed synQUAD dispenser powered by synQUAD liquid dispensing technology is the enabler for the CellJet printer to fabricate 3D cell constructs, which will be presented in the proceeding sections. Digilab also developed a piece of Windows-based control software, entitled Axsys, for automated low volume liquid handling applications. This user-friendly software (see Fig. 3.24) allows users to programme printing protocols with a graphical format.

Table 3.9. The CellJet printer specifications (courtesy of Digilab Inc.).

Specification	CellJet cell printer
Dynamic dispense range	20 nL to 1 mL (with 190 μm ceramic tip, 1 mL syringe and 1000 μl loop)
Dispense speed	30 seconds to dispense to a 96-well plate with a 1-channel system Dispense and aspirate speeds can be increased or decreased dependent upon cell type, and multiple channels can fill an entire plate faster
Positioning performance	Stepper motor resolution – 1.3 μm Repeatability < ± 10 μm
Control interface	External PC and CellJet software

Bioprinting

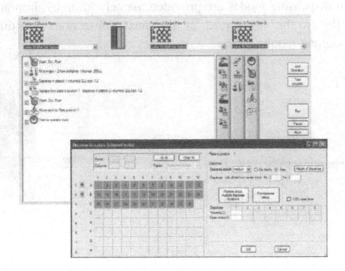

Fig. 3.24. The CellJet software interface (courtesy of Digilab Inc.).

3.9.3   *Process and Principle: Digilab's synQUAD Technology*

The synQUAD dispenser using synQUAD liquid dispensing technology is the primary function element that realises cell deposition. Thus, this subsection presents the working process and principle of synQUAD technology.

The synQUAD technology utilises a solenoid valve and a syringe stepper motor with a high resolution movement capability to control the number of drops to be released from a ceramic tip (see Fig. 3.25). A system fluid fills all the syringe pumps and lines providing a non-compressible environment. The syringe plunger is controlled and driven by the syringe stepper motor and the liquid is pushed into the solenoid valve. The pressure is generated within the system, which actuates the valve to enable drop acceleration and release onto the target. In order to have better control over the entire dispensing process and achieve better results, the valve should be placed as close as possible to the dispense tip.

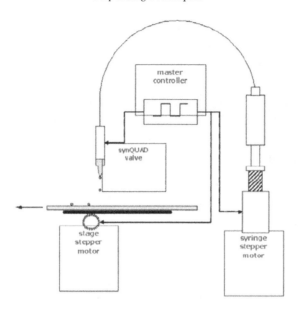

Fig. 3.25. Schematic of the platform mainly consisting of valve, syringe and stage synchronised by the synQUAD technology (courtesy of Digilab Inc.).

3.9.3.1 *Dispense modes*

There are two dispense modes available for the cell dispenser, which are Aspirate/Dispense and Continuous (bulk) Dispense Transfer. In the first mode, a sample from a source is aspirated and transferred to the defined area of the plate. For the continuous dispensing mode, the system is filled with the reagent and will only be used to dispense. This mode is for high throughput production where plates are quickly filled in or liquids are dispensed onto other devices.

3.9.3.2 *On-the-fly and stop-and-drop*

'On-the-fly' dispensing means high speed is the priority and a single channel can fill a 1536-well plate in less than 120 seconds. 'On-the-fly' dispensing is realised by synchronising the valve, syringe and the platform stage depicted in Fig. 3.25 above. The syringe displaces the desired drop volume when it moves across the plate. The solenoid is immediately actuated to eject the drop while the tip is accurately positioned over the target.

For certain applications where high speed is not necessary but positioning accuracy is the important factor to be considered, the 'stop-and-drop' dispense mode is an ideal candidate. A dispense tip can be placed in less than a millimetre above the destination. The resolution of the X and Y motors is 1.5 µm and the positioning accuracy is ± 10 µm. The 'stop-and-drop' motion can generate arrays within the well of a 96-well plate.

3.9.4 *Strengths and weaknesses*

The major benefits of the CellJet system are outlined below and the drawbacks of the CellJet system are similar to inkjek printheads presented in section 3.6.4.

(1) Fine volumes of droplets ranging from 20 nL to 4µL can be printed.
(2) High printing accuracy and throughput.
(3) High cell-viability of delicate and robust cells alike.
(4) synQUAD technology is capable of dispensing both low and high viscous fluids. The viscosity of the fluid (media/solution) can be as low as that of water e.g. typical cell media, or as high as that of 1-2% Alginate or 30% Polyethylene glycol.
(5) System dispenses a full range of cell types, from robust to delicate, both adherent and suspension cell types.

3.10 Valve-Based Printing II

3.10.1 *Company: nScrypt*

nScrypt, Inc. was founded in October 2002 as the result of a joint venture between Sciperio, specialising in cross-disciplinary solutions, and Spectra Technologies (now a part of Metro Automation, Inc.), an automation design and services company. Initially, nScrypt's role was to develop, manufacture, sell and service Sciperio's Micro Dispense Direct

Write (MDDW) technology [36]. This technology was then successfully applied to various material deposition applications in 2007. nScrypt's printing and micro dispensing technologies has enabled the 3D bioprinting of human tissue scaffolds. The company now manufactures micro dispensing pumps and systems for a wide range of materials. nScrypt is currently investing in a new frontier technology, aiming to construct the entire human tissue and organs. nScrypt Inc. is located at 12151 Research Parkway, Suite 150, Orlando, FL 32826, USA.

3.10.2 *Products: Tabletop and 300 Series Printers*

Currently, there are two models for scaffold printing, namely, the Tabletop and 3D-300 Series (see Fig. 3.26). The Tabletop Series micro dispense pump dispensing systems offer high precision in both 3D printing accuracy and dispensing volume due to the high motion system with the accuracy of ±10 μm and repeatability of ±2 μm. They are also flexible for future applications as the major elements i.e. dispense valves, vision, lighting, and graphical user interface as well as parts of the motion platform can be changed to meet new requirements. The working volume of a Tabletop machine is 304 × 152 × 101 mm (12 × 6 × 4") in the X, Y and Z axis. The 3D-300 machines are equipped with the nScrypt developed mapping technologies for non-uniform surfaces, which allow material to be dispensed accurately and consistently in the Z direction. nScrypt has specially developed a deposition tool named BioAssembly Tool (BAT), utilising CAD/CAM approach to build three-dimensional heterogeneous tissue models [37]. The BAT is a multihead, through-nozzle deposition tool used to conformably deposit biomaterials and cofactors on various supporting surfaces to create surrogate tissues and tentative platforms. The device contains an XY coordinate system with a stage; a number of Z-travelling deposition heads; a fiberoptic light source to illuminate the deposition area and cure photopolymers inline; individual ferroelectric temperature controls for each deposition head; a water jacket temperature control for the stage; and a piezoelectric humidifier [38]. The available biomaterials for the Tabletop and 300 Series printers are: poly(caprolactone) (PCL), collagen type I, hyaluronic acid (HA), poly(ethylene oxide) (PEO), poly(ethylene glycol) (PEG), fibrinogen, thrombin, calfskin gelatin and etc. nScrypt's printers are equipped with the Computer Aided Biology (CAB) technology, which was designed based on the CAD environment. The software is able to control the fabrication processes. It provides a graphical interface, which allows users to design, view and modify 3D models.

Fig. 3.26. Tabletop (left) and 300 Series machines (right) (courtesy of nScrypt Inc.).

The most important component that determines deposition accuracy is the patented Smart Pump™. The SmartPump™ controls, starts and stops material flow for an extreme range of viscosities. It uses positive pressure and a computer controlled needle valve. The computer controlled needle valve provides active valving and suck-back to control flow characteristics during start, stop and a wide range of other direct print scenarios. Depending on the pump chosen, the dead volume inside the SmartPump™ ranges from 0.025 to 0.1 cubic-centimetre. With these features and degrees of control, it is possible to dispense very small volumes of material (i.e. down to 20 picoliters) and a range of viscosities from 1 centipoise to over 1 million centipoise.

3.10.3 *Process and principle*

The printing process is controlled by specific software where parameters can be flexibly altered such as linear speed of deposition, deposition toolpath, air pressure and syringe plunger rate in the displacement heads. The reproducibility of the positioning of the extrusion nozzles can achieve a resolution of less than 5 μm. The extruder is pressure-driven and deposition rates are in the range of 12 nL/s to 1 mL/s. BAT versions are encapsulated in positively ventilated and positively pressurised casings. Optical and piezoelectric sensors enable material to be deposited onto curved surfaces.

3.11 Laser Printing

As non-contact and sterile techniques, optical methods have gained significant interest in the area of cell micromanipulation [39]. Laser printing is the dominant technique in the optical methods, which is the focus of this section.

3.11.1 *Laser Guidance Direct Write (LGDW)*

The laser guidance technique was first proposed by Odde and Renn [40, 41], which utilises radiation pressure to control and guide particle deposition. A typical configuration of LGDW, as shown in Fig. 3.27, is comprised of a weakly focused laser beam, a receiving substrate and particle suspension (liquid or aerosol) as the deposition materials. Functional optical forces with two components acting on particles (a radial and an axial component with different directions) are induced by using the weakly focused laser. Thus, particles can be pulled towards the centre of the laser beam (by the radial force) and pushed along the propagation of the light (by the axial force) [42]. As the force (normally > 10 pN) is much larger than the mass of a particle, the particles can be guided over distances up to a few millimetres. The size of a biological cell typically ranges from nano to micro-scale. As a result, cells can be deposited onto the surface to create constructs with required shapes. The major processing stage of LGDW is atomisation, during which droplets to be deposited onto the substrate are generated using one of the two methods, namely, nebulisation or sonic agitation. After droplets are formed, they will be guided onto the receiving substrate by either laser-momentum effects or gas flow [25].

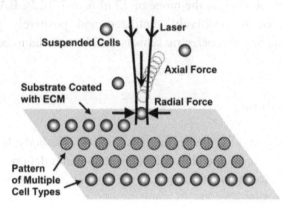

Fig. 3.27. Schematic of the laser guidance direct write [42] (Copyright 2006, with permission of WILEY-VCH Verlag GmbH & Co. KGaA, Weinheim).

3.11.2 *Laser Induced Forward Transfer (LIFT)*

LIFT, also known as biological laser printing (BioLP^{TM}), is originally developed by Barron *et al.* [43] in 2004. Since then, a number of improvements have been made and LIFT is thus regarded as modified-LIFT.

LIFT utilises a number of continuous focused laser pulses to move material onto a receiving substrate from a carrier support consisting of an optically transparent quartz disk (see Fig. 3.28). The quartz disk, coated with a metal oxide of 1-100 nm in thickness on the surface, functions as a laser absorption layer, which facilitates the laser pulses to be focused on its interface. A 10-100 μm thick biomaterial layer (solid, liquid, gel or powders) is coated above the absorption layer. The interaction between the laser and material enables photo-absorption of the laser energy, during which each individual laser pulse leads to the removal of the material from the carrier support, transferring an aliquot of the material travelling through air to the substrate.

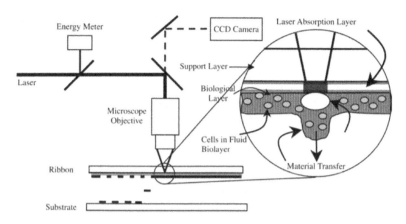

Fig. 3.28. Schematic of the LIFT process [43] (copyright 2004, with permission of Springer).

3.12 Electrohydrodynamic Jetting (EHDJ) Technology

Electrohydrodynamic is a physical phenomenon whereby a liquid medium is charged and fragmented [44]. The first success of using this technology to deposit living cells in suspension is reported by Jayasinghe *et al.* [45]. Recently, EHDJ, also known as electrospraying, is fast becoming a useful method for biotechnology, capable of processing a wide range of cells and biomaterials.

In an EHDJ process, cells are contained in a liquid medium which is charged and subsequently dispersed into the high intensity electric field of 0.85 kVmm^{-1} generated between the ground electrode and the needle (positive potential). In the electric field, various liquid shapes are formed and as a result, cell jets evolve. It should be noted that these jets are not the final droplets to be ejected. The jets will then be subjected to a number of reactions, which stimulates fragmentation and originates the formation of the final droplets. The major elements of a typical EHDL machine are a syringe pump, a ground electrode and a needle. The pump provides the kinematic energy, continuously transferring cell solution to the needle. The ejection of cell solution is realised through increasing electric field strength. This printing process is illustrated in Fig. 3.29 and a periodic spark can be visibly observed between the electrode and the needle as shown in Fig. 3.30.

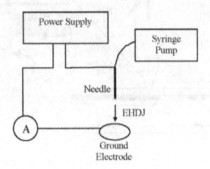

Fig. 3.29. Schematic of the EHDJ apparatus [25] (Copyright 2006, with permission of John Wiley and Sons).

Fig. 3.30. A discharging spark [45] (Copyright 2006, with permission of John Wiley and Sons).

The capabilities of the inkjet, laser and EHDJ techniques are compared in Table 3.10.

Table 3.10. Single lined table captions are centered to the table width. Long captions are justified to the table width manually.

	Print speed	Maximum cell throughput (cell/s)	Resolution (μm)	Cell viability (%)
Inkjet				
- thermal	5×10^3 (drops/s)	850	> 300	75 – 90
- piezoelectric	1×10^4 (drops/s)	2	5 – 254	–
LGDW	Continuous 9×10^{-8} (mL/s)	0.04	10 – 30	–
LIFT	1×10^2 (drops/s)	10^4	30 – 100	95 – 100
EHDJ	Continuous 0.01 (mL/s)	2×10^4	50 – 1000	–

3.13 Examples

3.13.1 *Organovo 3D Bioprinted Human Liver Tissue*

This subsection shows a real example of human liver tissue that was directly printed from Organovo's NovoGen MMX Bioprinter™. Fig. 3.31 is a cross section view of this 3D bioprinted liver tissue using three different cell types, namely primary liver cells (purple), stellate cells (green) and endothelial cells (red).

3.13.2 *Digilab Bioprinted Human Mesenchymal Stem Cells (hMSCs)*

Human Mesenchymal Stem Cells (hMSCs) are multipotent progenitor cells that can differentiate into a variety of tissue types and therefore greatly boost the regenerative capacity of the recipient tissue after transplantation [46]. Figure 3.32 shows the bioprinted hMSCs in 0.5% Sodium Alginate. The Digilab's CellJet printer was used to print the geometrical patterns in two 6-well plates.

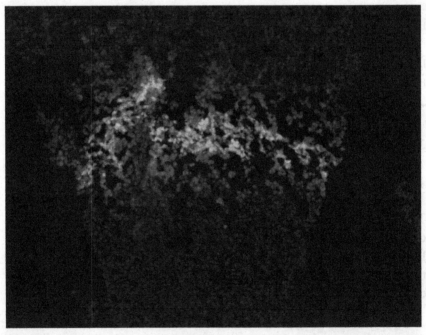

Fig. 3.31. Cross section of 3D bioprinted human liver tissue using 3 cell types (courtesy of Organovo Inc.).

Fig. 3.32. Bioprinted Human Mesenchymal Stem Cells in 0.5% Sodium Alginate in simple geometrical patterns in (A) sterile pre-wet 6-well plate; (B) or dry 6-well plate (courtesy of Digilab Inc.).

References

[1]   V. Mironov, N. Reis, and B. Derby, "Review: bioprinting: a beginning," *Tissue engineering,* vol. 12, pp. 631-634, 2006.

[2]   V. Mironov, R. P. Visconti, V. Kasyanov, G. Forgacs, C. J. Drake, and R. R. Markwald, "Organ printing: tissue spheroids as building blocks," *Biomaterials,* vol. 30, pp. 2164-2174, 2009.

[3]   C. K. Chua and K. F. Leong, *3D Printing and Additive Manufacturing: Principles and Applications*: World Scientific Publishing Company Inc., 2014.

[4]   EnvisionTEC. (2013, 29/11/2013). *3D Bioplotter*. Available:
        http://envisiontec.com/products/3d-bioplotter/
[5]   EnvisionTEC. (2013, 29/11/2013). *3D-Bioplotter® 4th
        Generation   Data   Sheet.*   Available:
        http://envisiontec.com/envisiontec/wp-
        content/uploads/2012/12/Machine-Bioplotter.pdf
[6]   EnvisionTEC. (2013, 29/11/2013). *Material Overview for the
        3D-Bioplotter®.*   Available:
        http://envisiontec.com/envisiontec/wp-
        content/uploads/2012/12/Bioplotter-Material-Overview.pdf
[7]   EnvisionTEC. (2013, 29/11/2013). *EnvisionTEC - Medical and
        Biofabrication.*   Available:
        http://envisiontec.com/envisiontec/wp-
        content/uploads/2012/12/Biofab-for-Web-041613.pdf
[8]   Organovo. (2013, 02/12/2013). *Changing the shape of medical
        research   and   practice.*   Available:
        http://www.organovo.com/company/history
[9]   L. De Bartolo and A. Bader, *Biomaterials for Stem Cell
        Therapy: State of Art and Vision for the Future*: CRC PressI Llc,
        2013.
[10]   F. Marga, K. Jakab, C. Khatiwala, B. Shephard, S. Dorfman, and
        G. Forgacs, "Organ printing: A novel tissue engineering
        paradigm," in *5th European Conference of the International
        Federation for Medical and Biological Engineering*, 2012, pp.
        27-30.
[11]   Organovo. (2013, 03/12/2013). *The bioprinting process.*
        Available: http://www.organovo.com/science-technology/
        bioprinting-process
[12]   C. Norotte, F. S. Marga, L. E. Niklason, and G. Forgacs,
        "Scaffold-free vascular tissue engineering using bioprinting,"
        Biomaterials, vol. 30, pp. 5910-5917, 2009.
[13]   S. Camazine, *Self-organization in biological systems*: Princeton
        University Press, 2003.
[14]   G. M. Whitesides and B. Grzybowski, "Self-assembly at all
        scales," *Science,* vol. 295, pp. 2418-2421, 2002.
[15]   D. Godt and U. Tepass, "Drosophila oocyte localization is
        mediated by differential cadherin-based adhesion," *Nature,* vol.
        395, pp. 387-391, 1998.

[16] J. M. Pérez Pomares and R. A. Foty, "Tissue fusion and cell sorting in embryonic development and disease: biomedical implications," *Bioessays,* vol. 28, pp. 809-821, 2006.

[17] R. A. Foty and M. S. Steinberg, "The differential adhesion hypothesis: a direct evaluation," *Developmental biology,* vol. 278, pp. 255-263, 2005.

[18] M. S. Steinberg, "Differential adhesion in morphogenesis: a modern view," *Current opinion in genetics & development,* vol. 17, pp. 281-286, 2007.

[19] P. A. Fleming, W. S. Argraves, C. Gentile, A. Neagu, G. Forgacs, and C. J. Drake, "Fusion of uniluminal vascular spheroids: a model for assembly of blood vessels," *Developmental Dynamics,* vol. 239, pp. 398-406, 2010.

[20] C. Khatiwala, R. Law, B. Shepherd, S. Dorfman, and M. Csete, "3D cell bioprinting for regenerative medicine research and therapies," *Gene Therapy and Regulation,* vol. 7, 2012.

[21] M. M. Mohebi and J. R. Evans, "A drop-on-demand ink-jet printer for combinatorial libraries and functionally graded ceramics," *Journal of combinatorial chemistry,* vol. 4, pp. 267-274, 2002.

[22] H. Sirringhaus, T. Kawase, R. Friend, T. Shimoda, M. Inbasekaran, W. Wu, *et al.,* "High-resolution inkjet printing of all-polymer transistor circuits," *Science,* vol. 290, pp. 2123-2126, 2000.

[23] T. Boland, T. Xu, B. Damon, and X. Cui, "Application of inkjet printing to tissue engineering," *Biotechnology journal,* vol. 1, pp. 910-917, 2006.

[24] Fujifilm. (2013, 15/12/2013). *Dimatix Materials Printer DMP-2831.* Available: http://www.fujifilmusa.com/products/industrial_inkjet_printhead s/deposition-products/dmp-2800/index.html

[25] B. R. Ringeisen, C. M. Othon, J. A. Barron, D. Young, and B. J. Spargo, "Jet-based methods to print living cells," *Biotechnology journal,* vol. 1, pp. 930-948, 2006.

[26] T. Xu, C. A. Gregory, P. Molnar, X. Cui, S. Jalota, S. B. Bhaduri, *et al.,* "Viability and electrophysiology of neural cell structures generated by the inkjet printing method," *Biomaterials,* vol. 27, pp. 3580-3588, 2006.

[27]   T. Cao, K. H. Ho, and S. H. Teoh, "Scaffold design and in vitro study of osteochondral coculture in a three-dimensional porous polycaprolactone scaffold fabricated by fused deposition modeling," *Tissue engineering,* vol. 9, pp. 103-112, 2003.

[28]   E. D. Miller, J. A. Phillippi, G. W. Fisher, P. G. Campbell, L. M. Walker, and L. E. Weiss, "Inkjet printing of growth factor concentration gradients and combinatorial arrays immobilized on biologically-relevant substrates," *Combinatorial Chemistry & High Throughput Screening,* vol. 12, pp. 604-618, 2009.

[29]   X. Cui and T. Boland, "Human microvasculature fabrication using thermal inkjet printing technology," *Biomaterials,* vol. 30, pp. 6221-6227, 2009.

[30]   E. Sachlos and J. Czernuszka, "Making tissue engineering scaffolds work. Review: the application of solid freeform fabrication technology to the production of tissue engineering scaffolds," *Eur Cell Mater,* vol. 5, pp. 39-40, 2003.

[31]   S. Parsa, M. Gupta, F. Loizeau, and K. C. Cheung, "Effects of surfactant and gentle agitation on inkjet dispensing of living cells," *Biofabrication,* vol. 2, p. 025003, 2010.

[32]   D. B. van Dam and C. Le Clerc, "Experimental study of the impact of an ink-jet printed droplet on a solid substrate," *Physics of Fluids,* vol. 16, p. 3403, 2004.

[33]   RegenHU. (2013, 31/12/2013). *RegenHU 3D Bio-Printers.* Available: http://www.regenhu.com/products/3d-bio-printing.html

[34]   Digilab. (2014, 02/01/2014). *synQUAD liquid dispensing technology.* Available: http://www.digilabglobal.com/synquad

[35]   Digilab. (2014, 02/01/2014). *CELLJET cell printer.* Available: http://www.digilabglobal.com/celljet

[36]   nScrypt. (2013, 28/11/2013). *High throughput manufacturing micro to pico-liter 3D conformal dispensing systems and laser micro machining tools customized for your needs.* Available: http://www.nscrypt.com/about-nscrypt/index.php

[37]   A. M. Kachurin, R. L. Stewart, K. H. Church, W. L. Warren, J. P. Fisher, A. G. Mikos, *et al.,* "Direct-write construction of tissue-engineered scaffolds," in *Materials Reseaerch Society Proceedings,* 2001.

[38]   C. M. Smith, A. L. Stone, R. L. Parkhill, R. L. Stewart, M. W. Simpkins, A. M. Kachurin, *et al.,* "Three-dimensional

bioassembly tool for generating viable tissue-engineered constructs," *Tissue engineering,* vol. 10, pp. 1566-1576, 2004.

[39]   B. Stuhrmann, H. G. Jahnke, M. Schmidt, K. Jahn, T. Betz, K. Muller, *et al.,* "Versatile optical manipulation system for inspection, laser processing, and isolation of individual living cells," *Review of scientific instruments,* vol. 77, pp. 063116-063116-11, 2006.

[40]   D. J. Odde and M. J. Renn, "Laser-guided direct writing for applications in biotechnology," *Trends in biotechnology,* vol. 17, pp. 385-389, 1999.

[41]   D. J. Odde and M. J. Renn, "Laser-guided direct writing of living cells," *Biotechnology and Bioengineering,* vol. 67, pp. 312-318, 2000.

[42]   R. K. Pirlo, D. Dean, D. R. Knapp, and B. Z. Gao, "Cell deposition system based on laser guidance," *Biotechnology journal,* vol. 1, pp. 1007-1013, 2006.

[43]   J. Barron, P. Wu, H. Ladouceur, and B. Ringeisen, "Biological laser printing: a novel technique for creating heterogeneous 3-dimensional cell patterns," *Biomedical microdevices,* vol. 6, pp. 139-147, 2004.

[44]   P. Eagles, A. Qureshi, and S. Jayasinghe, "Electrohydrodynamic jetting of mouse neuronal cells," *Biochem. J,* vol. 394, pp. 375-378, 2006.

[45]   S. N. Jayasinghe, A. N. Qureshi, and P. A. Eagles, "Electrohydrodynamic Jet Processing: An Advanced Electric-Field-Driven Jetting Phenomenon for Processing Living Cells," *Small,* vol. 2, pp. 216-219, 2006.

[46]   M. F. Pittenger, A. M. Mackay, S. C. Beck, R. K. Jaiswal, R. Douglas, J. D. Mosca, *et al.,* "Multilineage potential of adult human mesenchymal stem cells," *science,* vol. 284, pp. 143-147, 1999.

Problems

1.   List at least five popular bioprinting methods.
2.   Describe the process flow of EnvisionTEC's extrusion process.
3.   What is the principal concept of Organovo's extrusion process?
4.   What are the advantages and disadvantages of Organovo's extrusion process?

5.   Describe the four major steps for extrusion-based bioprinting process, from cell preparation to final product.

6.   Describe the process flow of Digilab's valve-based process.

7.   What are the printing principles of valve-based processes?

8.   Describe the process flow of Fujifilm's inkjet printing process.

9.   Compare the differences between extrusion-based and inkjet printing processes. What are the pros and cons?

10. What are the two important modes in Microjet's piezoelectric process? Describe their printing principles.

11. What is the major feature of RegenHU's BioFactory® that is different from other companies' products?

12. Compare RegenHU's BioFactory® with Digilab's CellJet. What are the advantages and disadvantages for each of the systems?

13. What are the two approaches in laser-based bioprinting technique? Describe their principles.

14. Using the Organovo's NovoGen MMX system as an example, in your opinion, what are the factors that limit the further development of the extrusion-based process?

Chapter 4

Material for Bioprinting

Biomaterials science and engineering has expanded, advanced and evolved dramatically since the late 1960s. It now covers fundamental aspects of physical, chemical, mechanical, electrical and even biological properties of both the natural and synthetic biomaterials. This chapter aims to introduce the most popular biomaterials in relation to tissue engineering (TE). Most of these biomaterials have been proven to be available for bioprinting. The requirements of biomaterials specifically in tissue engineering are first presented, followed by polymers, ceramics and glasses. Finally, this chapter pays particular attention to hydrogels and their unique properties.

4.1 Overview of Biomaterials

4.1.1 *Overview of biomaterials and its definition*

Biomaterials are non-viable materials typically used in therapeutic and diagnostic systems that are in contact with tissue or biological fluids [1]. Biomaterials can be categorised into polymers (natural and synthetic), ceramics, metals (alloys), glasses, carbons and composites comprised of various combinations of the above material types [2]. Biomaterials have been developed with the purpose of replacing the function of the biological materials. With the unique and superior material properties, biomaterials have been widely used to fabricate a large range of medical devices, pharmaceutical preparations as well as diagnostic products in medical and healthcare applications. Given that biomaterials are used in intracorporeal environments, they must be:

- Nontoxic and noncarcinogenic.
- Chemically stable and resistant to corrosion, able to sustain large and variable stresses in the human body.
- Able to be shaped/manufactured into intricate geometries.

4.1.2 Requirements of biomaterials in tissue engineering

4.1.2.1 Formability

Bioprinting techniques are highly specialised technologies in terms of material formability. Each technique requires a specific form of input material such as liquid suspension. As a result, it should always be ensured that the choice of materials is compatible with the bioprinting process to be used so that the selected material can be efficiently produced in the desired form.

4.1.2.2 Water content

The amount of water in a hydrogel directly determines the absorption rate and the solute diffusion through the hydrogel. The water content in the hydrogel should be well controlled if the content is higher than necessary which may lead to deterioration of cell proliferation rate [3]. Water may be bound onto the gel by two interactions, primarily via hydration of the polar hydrophilic groups [4]. Additional water is absorbed as free water that fills the space such as voids and macropores formed from between the chains. Water content in the hydrogel can be determined either by identifying the light absorbance in the gel or measuring the percentage change in the mass between the hydrogel and its dry form [5, 6].

4.1.2.3 Biocompatibility

Biocompatibility is a critical property for a biomaterial. The definition has evolved along with the advances in materials used in medical applications [7]. There are two definitions that have been widely adopted, which are:

- '*the quality of not having toxic or injurious effects on biological systems*' [8].
- '*the ability of a material to perform with an appropriate host response in a specific application*' [9].

There are three points that should be addressed to further describe the biomaterial definition:

(1) Biocompatibility is a collection of processes that include a number of interdependent mechanisms of interaction taking place between the tissue and the material. Biocompatibility, though an important material property, cannot be considered as an intrinsic property. The biocompatibility of a material refers to a specific application in which the material is used. No material is definitely biocompatible. Despite the fact that a number of materials may be biocompatible under one or more conditions, it cannot be considered that they are able to present biocompatibility in all conditions

(2) The material implanted in a human body is expected to perform a specific function as opposed to staying there. Therefore, biocompatibility is also considered as the ability of the material to continuously perform a function.

(3) Appropriate/acceptable host responses are allowed, which, in other words, indicate that a biocompatible material is not necessarily required to generate no response.

4.1.2.4 *Suitable mechanical properties*

High static and cycle-dependent properties are desired properties that metals should have. The primary metallic strength attributes include tensile yield, modulus of elasticity, ultimate strength and fatigue endurance. Other mechanical properties should also be addressed for specific applications such as creep and compressive yield strengths for dental applications. In terms of polymers, principal mechanical properties are tensile, fatigue and creep strengths as well as modulus. In addition, since excessive wear can result in premature mechanical failure of the implants leading to wear debris that may not be biocompatible to the host, wear resistance is one of the priorities when choosing biomaterials.

4.1.2.5 *Biodegradability*

The degradation properties of a biomaterial are critical for the success of the scaffold-based bioprinting approach. This is because in an ideal scenario, the scaffold made of the biomaterial will be remodelled as well as resorbed by growing cells and gradually replaced by the newly formed differentiated cells and extracellular matrix (ECM). A desirable feature is the synchronisation of the polymer degradation rate with the rate of tissue ingrowth [10]. The degradation-absorption mechanism is caused by a number of interrelated factors including degree of crystallinitya, the hydrophilicity of the polymer backbone, volume of porosity and the surface area as well as presence of catalysts. Balancing each of these factors properly will facilitate an implant to degrade slowly whilst transferring stress at an appropriate rate to the surrounding tissues as they heal.

4.1.2.6 *Biodegradation product*

Although degradation products of biodegradable polymers are known as largely non-cytotoxic, little information is available with respect to the degradation rate-dependent acidic byproduct effect of the biomaterial. Sunga *et al.* [11] reported that fast degradation of the polymer leads to detrimental impact on cell viability and migration into the scaffold, both *in vitro* and *in vivo*. A widely accepted explanation is the rapid local acidification due to polymer degradation.

4.1.2.7 *Bioactivity*

What is bioactivity? Prior to answering this question, the term 'bioactive material' should be introduced. In general, 'bioactive' signifies that the material inspires a positive and advantageous biological response from the body where the implant resides. Bioactive was first proposed by

---

[a] A measurement of how much of the polymer is incorporated into crystalline regions

Larry Hench [12] in 1971 who invented the first material that was able to form a strong bond to bone. Therefore, bioactivity initially represented materials that could bond to bone. Later, this definition was expanded and materials that can bond to soft tissues or release biological stimulants are also considered as bioactive.

The interaction between material and cells is controlled by both chemical and structural signalling molecules that play a decisive role for cell adhesion and further cell behaviour after initial contact [13]. The extent of initial cell adhesion determines the shape, size, number and distribution of focal adhesion plaques formed on the cell membrane, which then defines the shape and size of the cell-spreading area. The extent of spreading is vital for further proliferation, migratory and differentiation behaviour of anchorage-dependent cells. The current strategies used to control the proliferation and other cell behaviours on biomaterials include patterning the material surfaces with adhesive molecules or by incorporating a controlled release of biomolecules, e.g. hormones, enzymes, natural growth factors or synthetic cell cycle regulators.

4.1.2.8 *Sterilisation considerations*

Sterilisation is an essential process for all materials and/or devices to be implanted. For economic considerations, expensive devices (e.g. surgical instruments) are used repetitively, in which cases, sterilisation is compulsory before the devices are implanted to other patients. There are three traditional sterilisation methods commonly used, including steam sterilisation, gamma radiation sterilisation and ethylene oxide (EtO) gas sterilisation [14]. Each method must achieve the same goal, namely, destroying or removing living organisms and viruses from the device/material. The selection of a sterilisation method depends on implant material properties (e.g. radiation sensitivity and heat resistance) and economic considerations. The sterility assurance limit (SAL) is a tool for quantifying sterility, which can be expressed as the probability that an implant will remain a nonsterile state following a sterilisation run. The accepted value for the SAL is less than 10^{-6}, representing that one implant in one million will possibly be nonsterile.

4.2 Polymers

A polymer is a large molecule consisting of numerous repeated subunits i.e. monomers. Owing to their wide range of properties, both natural and synthetic polymeric products have become essential and ubiquitous daily supplies in our lives. Polymers ranging from natural biopolymers such as proteins to synthetic plastics, polystyrene for instance, are fundamental to biological structures and functions. By adjusting combination ratios during polymerisation, unique physical properties such as viscoelasticity and toughness can be obtained. This section is focused on synthetic polymers, introducing their properties and applications.

4.2.1 *Natural polymers*

Natural polymers that are used in bioprinting and tissue engineering are normally hydrogels. Therefore, the natural polymers including gelatin, collagen, hyaluronic acid (HA), fibrin, alginate, chitosan and chitin will be presented in section 4.4 (Hydrogel).

4.2.2 *Synthetic polymers*

4.2.2.1 *Poly-L-lactic acid (PLLA)*

Poly-L-lactic acid is a synthetic resorbable and biodegradable polymer, which falls into the alpha-hydroxy-acid group of compounds. It is a crystalline polymer with approximately 37% crystallinity and the crystallinity degree is dependent on the polymer processing parameters and the molecular weight. The glass transition and melting temperatures are normally 60-65°C and 175°C respectively [15]. In contrast to polyglycolide, PLLA is a slow-degrading polymer which exhibits low extension, high tensile strength and a high modulus (~4.8 GPa), making it a suitable candidate for load bearing. PLLA can be used to form high strength fibres and thus has been used in the fabrication of scaffolds for ligament replacement [16, 17]. However, 2 to 5.6 years are usually needed to fully degrade PLLA with high molecular weight *in vivo* [18].

In addition, the degradation rate is related to the degree of crystallinity and the matrix porosity.

PLLA has been used since the 1990s in various orthopaedic and maxillofacial procedures [19]. In 1999, implants that contained PLLA for intradermal injection were approved in Europe for the correction of skin depression and the increase of the volume of depressed skin areas such as scars, wrinkles and skin creases. PLLA was also approved in the US for the correction and restoration of the signs of facial fat loss due to severe facial lipoatrophy induced by human immunodeficiency virus (HIV) infection.

4.2.2.2 *Poly(glycolic acid) (PGA)*

Poly(glycolic acid) is a thermoplastic with high rigidity and crystallinity (46-50%). This high crystallinity makes it insoluble in most organic solvents except highly fluorinated organic solvents e.g. hexafluoro isopropanol. The PGA molecule is presented in Fig. 4.1. The glass transition (T_g) and melting temperatures (T_m) are 36°C and 225°C respectively. Due to the degradation product (glycolic acid) being a natural metabolite, PGA has become very popular in medication application such as resorbable sutures. Porous scaffolds and foams can also be manufactured from PGA. However, it should be noted that the properties and degradation characteristics may vary when using different processing techniques.

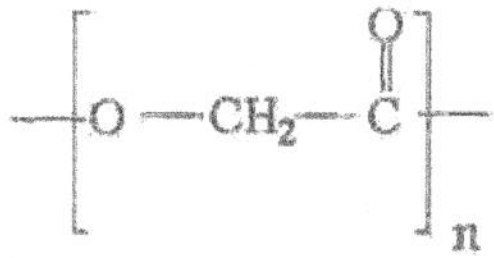

$$\left[O - CH_2 - \overset{\displaystyle O}{\overset{\displaystyle \|}{C}} \right]_n$$

Fig. 4.1. Molecular structure of poly(glycolic acid).

In general, the degradation of PGA involves random hydrolysis of its ester bonds. Apart from hydrolysis, PGA is also decomposed by certain enzymes, in particular those with esterase activity [20]. Glycolic acid can be excreted by urine. The degradation rate is determined by a number of factors including configurational structure, crystallinity, morphology, molecular weight, site of implantation and etc. Both *in vitro* and *in vivo* studies have suggested that PGA is sufficiently biocompatible [21, 22]. However, two major concerns have also been raised. The first one lies in orthopaedic applications which normally require implants with large size, leading to release of degradation products containing high local acid concentrations. It was identified that PGA could produce toxic solutions during acidic degradation [23]. Another concern is the inflammatory response which can be triggered by small particles released during degradation. As PGA has successfully been applied to clinical applications such as sutures, their use in fixation devices or replacement implants are considered to be biocompatible.

4.2.2.3 *Poly(caprolactone) (PCL)*

Poly(caprolactone) is a semicrystalline polymer, of which the T_g and T_m are approximately -60°C and 59-64°C respectively. The molecular structure of PCL is shown in Fig. 4.2. PCL is recognised as a non-toxic and tissue compatible material [24]. It degrades at a low rate and typically a molecular weight of 50,000 takes nearly three years to degrade completely *in vitro*. PCL can be used as a base polymer in development of long term, implantable drug delivery systems. It is prepared by the ring-opening polymerisation of the cyclic monomer ε-caprolactone [25].

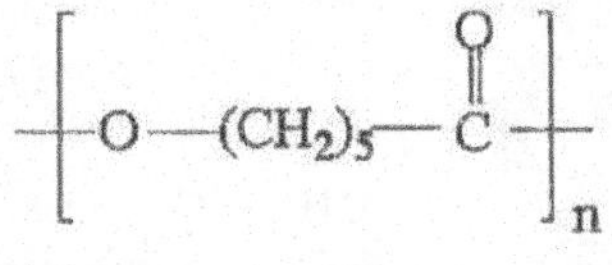

$$\left[O-(CH_2)_5-\overset{\overset{\textstyle O}{\|}}{C} \right]_n$$

Fig. 4.2. Molecular structure of poly(caprolactone).

4.2.2.4 *Poly(lactide-co-glycolide) (PLGA)*

L- and DL-lactides are both used for co-polymerisation. poly(L-lactide-co-glycolide) becomes amorphous polymers when using the composition range of 25-75%. 50/50 PLGA is highly hydrolytically unstable and at either end of the co-polymer composition range, PLGA exhibits high resistance to hydrolytic degradation [26, 27]. PLGA undergoes bulk erosion through ester bond hydrolysis. The degradation rate is affected by various parameters such as the LA/GA ratio, matrix structure and molecular weight [28]. PLGA is a very popular co-polymer due to its good processibility which facilitates various structures and forms to be fabricated. It has been used to fabricate scaffolds and control drug delivery systems.

4.3 Ceramics and Glasses

Ceramics and glasses involve a wide range of inorganic/nonmetallic compositions. In this section, the major types of ceramic and glass are introduced, focusing on their unique properties associated with tissue engineering.

4.3.1 *Hydroxyapatite (HAP)*

Synthetic hydroxyapatite is a popular material for bone repair as it is chemically similar to bone mineral [29]. Synthetic HAP can be produced by solution chemistry or obtained from natural materials. Synthetic HAP is considered as osteoconductive since it facilitates bone cell attachment, migration, proliferation and phenotypic expression, resulting in formation of new bone [30].

It is obvious that the microstructure can be altered by changing the porosity which also determines the mechanical strength of HAP ceramics [31]. When increasing preparation temperature, the porosity and surface area of HAP ceramics decrease accordingly. It is also noted that increasing porosity results in decreased compressive strength. Therefore both density and porosity should be addressed when choosing HAP ceramics. Table 4.1 lists the key mechanical properties of dense and porous HAP ceramics and adult human cortical bones. HAP ceramics have been extensively used in bone, soft tissues as well as other organs due to their excellent biocompatibility [32]. Ceramic implants can slowly dissolve and thus be replaced by native bone, by which the mechanical properties of the tissues are maximised.

Table 4.1. Mechanical properties of HAP
ceramics and adult human cortical bones
(adapted from [33]).

Property	HAP ceramics	Cortical bone
Mechanical strength (Compressive MPa)	7–69(p), 207–897(d)	~140
Tensile strength (MPa)	2.5(p), 19–193(d)	~70
Elastic modulus (GPa)	34–117(p)	3–20
Ultimate strength (MPa)	13(p)	130
Ultimate strain	~0.20%(d)	~0.02%
Ultimate stress – tension	~125	~130

(d) – dense, (p) – porous

4.3.2 *Alumina*

Alumina has been a very popular material for producing components of surgical and prostheses devices over the past four decades. Alumina is an inert material and highly resistant to corrosion. Hence, it can remain stable for a long service, usually more than 10 years, during which, minimal response from the tissues is induced. It exists in seven crystal phases, namely, α, δ, γ, η, θ, and χ. The most common alumina is α-alumina, which is a nonporous, dense and nearly inert material. Its hardness is close to diamond (9 on the Mohs scale). The material properties (e.g. strength, fracture and fatigue resistance) are dependent on the grain size, material purity and porosity. Table 4.2 shows its major mechanical properties.

Table 4.2. Mechanical properties of 99.5% pure
alumina (adapted from [33]).

Density	3.97 gm/cm^3
Flexural strength	345 MPa
Compressive strength	2100 MPa
Poisson's ratio	0.21
Fracture toughness	3.5 MPa.m$^{1/2}$
Hardness	1000 kg/mm^2
Elastic modulus	300 GPa
Shear modulus	124 GPa
Bulk modulus	172 GPa

Alumina is biocompatible but not bioresorbable, which means that it is recognised as a foreign material in the human body. The resulting action of the body is to isolate the alumina implant by generating a fibrous capsule wrapping around it. Although several solutions are available to considerably prevent formation of such a fibrous capsule, foreign-body reactions are unavoidably induced by the alumina particles generated during the wear of the alumina implant.

4.3.3 *Bioactive glass*

Bioactive glasses are ideal candidates for tissue engineering as they are amorphous silicate-based materials which bond to bone and can stimulate new bone growth while dissolving over time [1]. 45S5 Bioglass® (composition: 46.1% SiO_2, 24.4% NaO, 26.9%CaO and 2.6% P_2O_5, in mol%) is the first bioactive glass that formed an interfacial bond with host tissue after it was implanted into the body [12]. It was also found that the strength of the interfacial bond between cortical bone and Bioglass® was even greater than the strength of the host bone [34]. Other glass types with differing compositions have also been identified as bioactive. In addition, phosphate-based glasses, which have a high rate of resorption, have also been developed.

4.3.4 *Clinical products*

4.3.4.1 *Hydroxyapatite*

A typical HAP clinical application is the coating of orthopaedic implants. Coatings are deposited using the plasma-spray method. This spraying process uses a source material of HAP which is a mixture of crystalline calcium phosphates containing 95% HAP, and an amorphous calcium phosphate [35]. The aim of using HAP coated implants is to form a strong bond between the host bone and the metal alloy whilst eliminating the need for bone cement. Another successful HAP product is ApaPore® (Apatech Ltd., Elstree, UK). It is a porous HAP that has an interconnected macroporosity and some microporosity. ApaPore® is used in impaction grafting for spinal fusions, bone defect treatment and the cemented revision of total joint arthroplasties.

4.3.4.2 *Alumina*

Owing to the superior wear resistance, alumina has been used in orthopaedic applications since the 1970s. Figure 4.3 shows an example of total knee replacement and hip replacement implants. Generally, a metallic femoral stem, an alumina femoral head and an acetabular cup made of polyethylene are used together for hip replacements [33].

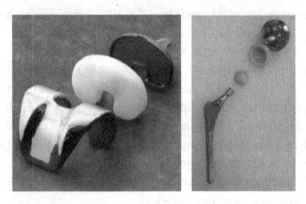

Fig. 4.3. Total knee (left) and total hip replacement (right) implants [33] (Copyright 2013, with permission of Elsevier).

Porous alumina has been used as bone spacers for replacing bone sections which have been removed owing to traumatic injury or cancer. Bone spacers are implanted into natural tissue. The porosity of alumina implants should be more than 30 % and typical pore sizes should not be smaller than 100 μm. This is to allow bone cell infiltration, encouraging vascularisation and enabling new bone formation.

For dental applications, dense alumina (single crystal) has been used as tooth replacements and some implants can be found in Fig. 4.4. However, the usage of alumina dental implants has continuously been declining recently due to their high elasticity modulus compared with the native tissue and inability to bend.

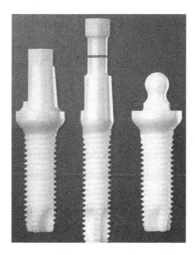

Fig. 4.4. Alumina dental implants [33] (Copyright 2013, with permission of Elsevier)

4.4 Hydrogels

Hydrogels have drawn significant attention in bioprinting and tissue engineering due to their high water contents. A hydrogel is actually a three-dimensional network of a series of hydrophilic polymer chains that are crosslinked through either physical bonding or chemical bonding, in the presence of water. In essence, hydrogels are water-swollen gels and of polymeric structures held together by: (1) primary covalent crosslinks; (2) hydrogen bonds; (3) ionic forces; (4) affinity or "bio-recognition" interactions; (5) physical entanglements of individual polymer chains; (6) polymer crystallites; (7) hydrophobic interactions; or (8) a combination of two or more of the above interactions. Hydrogels are highly absorbent natural or synthetic polymers and they can contain water of up to 99.9%. Some chemical hydrogels can even reversibly swell or shrink up to 1000 times in volume because of small changes in their environment such as temperature, pH and electric field.

There are a number of ways to categorise hydrogels, based on preparation methods, ionic charge or physico-chemical structural features. In this section, hydrogels are simply classified as natural and synthetic and the typical hydrogels are introduced, followed by their unique properties in bioprinting.

4.4.1 Natural polymers

4.4.1.1 Collagen

Collagen, a well-known protein, has been extensively used in biomedical applications as it is the main component of natural ECM and also the most abundant protein in mammalian tissues [36]. Although a large number of natural and synthetic polymers are used as biomaterials, the characteristics of collagen are distinct from those of synthetic polymers largely in its mode of interaction in the body. More than 20 collagen types have been found but their basic structures are the same, consisting of three polypeptide chains. These three chains wrap around one another, thereby creating a three-stranded rope structure. The strands are held together by covalent bonds and hydrogen. Stable collagen fibres can initiatively form as a result of strands self-aggregating. Furthermore, the mechanical properties of collagen fibres can be enhanced by introducing chemical crosslinkers such as carbodiimide and glutaraldehyde [37, 38]. Collagen can be degraded naturally by metalloproteases and thus, the degradation process can be locally controlled by cells in the engineered tissue.

4.4.1.2 Gelatin

Gelatin, a protein-based polymer, is derived through partial hydrolysis of collagen. Chemical pre-treatment followed by heat treatment disorganises collagen protein structure resulting in helix-to-coil transition and conversion into soluble gelatin. Gelatin has a lower antigenicity as compared to collagen [33]. Gelatin undergoes gelation during a change in temperature. However, gelatin has been modified to photopolymerisable hydrogel by addition of methylacrylate group (methylacrylaed gelatine or GelMA). Several studies have used GelMA for bioprinting [39-42].

4.4.1.3 *Fibrin*

Fibrin has recently been used as injectable scaffolds and cell delivery vehicles [43]. The major advantage is fibrinogen can be autologously derived from plasma, which reduces the risks induced by a foreign body reaction. Fibrin is generally used as glue consisting of thrombin and fibrinogen solutions. In surgery, fibrin glue is primarily employed to control bleeding. In addition, it has also shown promise in skin grafts and exogenous growth factor delivery by significantly reducing wound healing time.

4.4.1.4 *Alginate*

Alginate has shown its value in a wide range of medical applications including drug stabilisation and delivery, and cell encapsulation. It has low toxicity and gels under gentle conditions, and can be derived from brown seaweed and bacteria. Alginate is a polysaccharide copolymer of α-L-guluronic acid (G) and (1–4)-linked β-D-mannuronic acid (M) monomers (see Fig. 4.5). The G and M monomers are distributed sequentially in either alternating or repeating blocks [44]. The species, location, and seaweed age are the influential factors that determine the amount and distribution of each monomer. Alginate gels are formed as a result of divalent cations (e.g. Ba^{2+} or Ca^{2+}) cooperatively interacting with G monomer blocks to generate ionic bridges between polymer chains. By changing and adjusting the G and M ratio together with molecular weight of the polymer chain, the crosslinking density and the resulting mechanical properties can be easily manipulated. It is noted that ionically crosslinked alginate hydrogels undergo slow and uncontrolled dissolution rather than following a specific degradation trend.

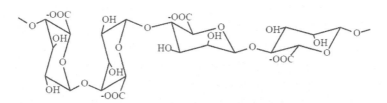

Fig. 4.5. Chemical structure of alginate.

4.4.1.5 *Chitin and Chitosan*

Chitin and chitosan have shown to be useful in wound dressing and drug delivery. Chitin is a polysaccharide consisting of randomly distributed *N*-acetyl-glucosamine and *N*-glucosamine units. The monomer units can also be block distributed depending on the derivation method [45]. Chitosan is a variant of chitin, a deacetylated counterpart. The biopolymer is termed chitin if the number of *N*-acetyl-glucosamine units is higher than 50%. Otherwise, the biopolymer is termed chitosan. Chitin and chitosan are commercially obtained from shellfish sources such as shrimps and crabs.

Chitosan is degradable by enzymes in humans and is structurally similar to glycosaminolglycans. Figure 4.6 depicts its molecular structure. Chitosan can easily be dissolved in dilute acids and can then be gelled by increasing the pH value. Chitosan is degraded by lysozyme and the degradation kinetics is inversely proportional to the degree of crystallinity.

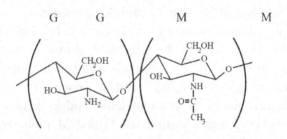

Fig. 4.6. Chemical structure of alginate.

4.4.1.6 *Hyaluronic acid*

HA is a glycosaminolglycan (GAG) and can be found in almost every mammalian tissue. It can be naturally degraded by hyaluronidase and has been widely used in wound healing and the synovial fluid of joints. HA is a linear polysaccharide consisting of a repeating disaccharide of (1–3) and (1–4)-linked β-D-glucuronic acid, and *N*-acetyl-β-dglucosamine units [46].Its molecular structure is shown in Fig. 4.7. A number of methods can be used to form HA hydrogels, such as covalent crosslinking with hydrazide derivatives, annealing and esterification. Moreover, HA can be combined with both alginate and collagen to form composite hydrogels.

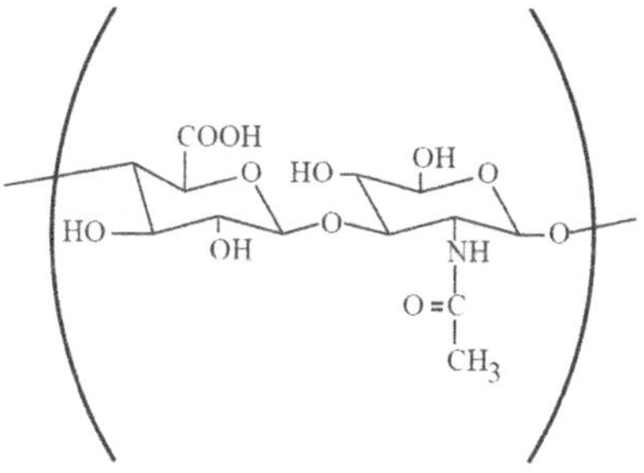

Fig. 4.7. Chemical structure of hyaluronic acid.

4.4.2 *Synthetic hydrogels*

The natural hydrogels presented in section 4.4.1 are increasingly popular due to the superior inherent biocompatibility. In contrast, synthetic hydrogels have shown their benefits in the regenerative medicine field, including highly tunable and consistent properties, and easier large-scale production. This subsection presents three most commonly used synthetic hydrogels.

4.4.2.1 *Poly(2-hydroxethyl methacrylate) (PHEMA)*

PHEMA hydrogels have been used as implant materials since the late 1960s. The gel networks can be formed by precipitation polymerisation of 2-hydroxyethyl methacrylate [47]. The resultant hydrogel is biologically inert but relatively weak. It also exhibits high resistance to cell adhesion and protein adsorption. However, it is found that PHEMA implants undergo calcification *in vivo* [48]. The molecular structure of neutrally charged PHEMA repeat units is shown in Fig. 4.8.

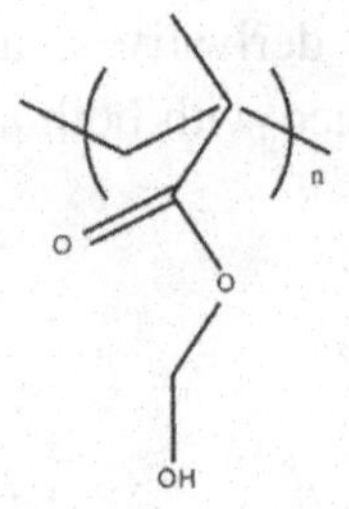

Fig. 4.8. Neutrally charged PHEMA monomers.

4.4.2.2   *Poly(vinyl alcohol) (PVA)*

PVA (see Fig. 4.9) is derived from hydrolysis of poly(vinyl acetate). In order to obtain a gel state, PVA is crosslinked through either chemical or physical methods. It can also be photo-cured to fabricate hydrogels. PVA is similar to PHEMA in the aspect of having available pendant alcohol groups that function as attachment sites for biological molecules [49]. PVA hydrogels are neutral and non-adhesive to proteins and cells. They have a low friction coefficient and the structural properties are similar to natural cartilage. More importantly, they are generally stronger than most other synthetic gels, which makes them successful for avascular tissue. Furthermore, PVA can be copolymerised with Poly(ethylene glycol) (PEG) to produce a biodegradable hydrogel which has a degradation rate faster than that of PEG hydrogels but slower than that of PVA homopolymer hydrogels.

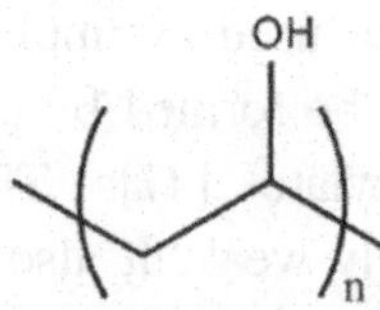

Fig. 4.9. Neutrally charged PVA monomers.

4.4.2.3 *Poly(ethylene glycol) (PEG)*

PEG (see Fig. 4.10) is a biocompatible and hydrophilic material and its homopolymer is a polyether which can be polymerised from ethylene oxide by condensation [50]. The hydroxyl groups on the PEG chains are frequently modified to produce a wide variety of derivatives. PEG is non-adsorptive due to the absence of protein binding sites. PEG hydrogels have been considered as one of the most successful synthetic hydrogels for TE applications. For instance, photocurable PEG hydrogels are extensively employed in cell encapsulation owing to its inert nature. Further, PEG can act as a mediator for immobilising the RGD sequence, a segment of protein that allows cells to attach on.

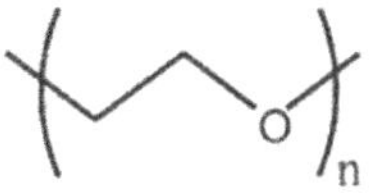

Fig. 4.10. Neutrally charged PEG monomers.

4.4.3 *Key hydrogel properties in bioprinting*

The suitability of a hydrogel for bioprinting is largely dependent on the physiochemical properties. The major physiochemical parameters that affect the hydrogel printability are the rheological properties and crosslinking mechanisms.

4.4.3.1 *Rheology*

Rheology is a discipline that mainly focuses on the flow of matter when applying an external force. In this section, the influence of some typical rheological parameters (i.e. viscosity, viscoelasticity, shear thinning and yield stress) is discussed.

- Viscosity

    Viscosity is the resistance of a fluid to flow under the application of stress. Fluids can be roughly classified into two types, namely,

Newtonian and Non-Newtonian fluids. The viscosity of a Newtonian fluid is dependent only on temperature but not on shear rate and time. Some examples of Newtonian fluids around us are water, milk, sugar solution and mineral oil. The viscosity of a Non-Newtonian fluid does not only depend on temperature but also on shear rate.

According to how viscosity changes with shear rate, the flow behaviour is characterised as shown in Fig. 4.11.

(i) Shear thinning: the viscosity decreases with increased shear rate. This will be further explained in the following section.

(ii) Shear thickening: the viscosity increases with increased shear rate.

(iii)Plastic: a fluid exhibits a so-called yield value; a certain shear stress must be applied before flow occurs.

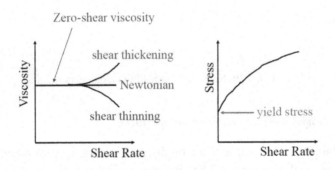

Fig. 4.11. Shear thinning, shear thickening and plastic characterisation.

There are two important properties for Non-Newtonian time dependent fluids, which are pheopexy and thixotropy. An increase in apparent viscosity with time under constant shear rate or shear stress, followed by a radical recovery when the stress or shear rate is removed [51]. This property is named pheopexy, anti-thixotropy or negative thixotropy. On the other hand, thixotropy property represents a decrease in apparent viscosity with time under constant shear rate or shear stress, followed by a gradual recovery, when the stress or shear rate is removed. Rheopectic and thixotropic properties under a constant shear rate is depicted in Fig. 4.12.

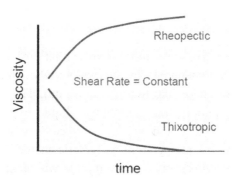

Fig. 4.12. Rheopectic and thixotropic properties of non-Newtonian fluids.

In bioprinting, a high viscosity is detrimental for surface tension-driven droplet formation. The molecular weight and polymer concentration are the two major factors that determine the viscosity of a polymer solution. The bioink viscosity directly affects shape fidelity and high viscosity is expected since high viscosity can lead to high printing fidelity. Nevertheless, it should be noted that the increased viscosity leads to an increase of the shear stress, resulting in the damage of the suspended cells. In addition, viscosity is somewhat affected by the solubility parameter, temperature, shear rate and other specific interactions. Table 4.3 shows the viscosities of some natural and synthetic hydrogel precursor solutions.

Table 4.3. Viscosities of hydrogel precursor solutions for bioprinting.

Hydrogel	Viscosity (Pa·s)	Concentration (% w/v)	Shear rate (s^{-1})	Molecular weight (kDa)	Reference
Alginate	0.9	2	100	100–500	[52]
Collagen type I	10	0.3	0.1–100	115–230	[53]
PEG	0.008	10	200–1300	3.35	[54]
HA	22	1.5	1	950	[55]

- Viscoelasticity

Viscoelastic materials are those whose relationship between stress and strain is time-dependent. In such materials, their stiffness will depend on the rate of loading (deformation rate). The time dependence leads to a behaviour that combines liquid-like and solid-like characteristics. Polymers and gels are typical viscoelastic materials. Depending on time or temperature, polymeric materials exhibit typical features of glass, rubber or viscous liquid, which give rise to wide applicability and good processability of polymers. If the deformation is small or applied sufficiently slowly, the molecular arrangements are never far from equilibrium. In this case, the viscoelasticity is considered to be linear. Otherwise, it is nonlinear.

- Shear thinning

Shear thinning relates to the non-Newtonian behaviour where increasing shear rate results in decrease of viscosity [56]. Shear thinning is more obvious for polymer solutions with high molecular weight. For instance, shear thinning for alginate is described by the decline in viscosity when increasing the shear rate. From Fig. 4.13, it can be identified that the viscosity decreases as the concentration of alginate increases.

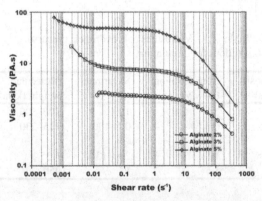

Fig. 4.13. Viscosity variation as a function of shear rate for different alginate solutions [57] (Copyright 2009, with permission of John Wiley and Sons).

- Yield stress

Yield stress is a stress that has to be overcome for initiating flow. In general, interactions between polymer chains leads to the formation of a fragile, physically crosslinked network. This network is broken by shear forces (above the yield stress) and gradually reforms after the shear forces are removed. In comparison to high viscosity which can only delay the collapsing process of a deposited 3D construct, the existence of a yield stress can prevent flow, collapse and cell settling in the hydrogel precursor reservoir. For instance, gellan gun is a typical anionic polysaccharide which can be crosslinked by cations to create physical networks [58]. By adding the gellan gun into gelatin methacrylamide (GelMA) at specific salt concentrations, it will form a gel with suitable properties for robotic dispensing due to the strong yield stress behaviour it exhibits. Figure 4.14 illustrates the effect of having both shear thinning and yield stress in GelMA and gellan gum. The gellan chains (white) form a provisional network and induce gel-like viscosity in the syringe (see Fig. 4.14(i)). Having dispensed the gel through a needle, the provisional network is broken up by shear whilst all polymer chains align, which significantly reduces the viscosity by orders of magnitude (see Fig. 4.14(ii)). The provisional network is regained immediately after the shear stress is removed and thus the plotted filament solidifies instantly, as shown in Fig. 4.14(iii).

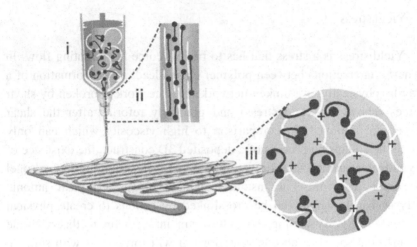

Fig. 4.14. Illustration of shear thinning and yield stress in GelMA and gellan gum [59] (Copyright 2013, with permission of John Wiley and Sons)

4.4.3.2 *Crosslinking mechanisms for hydrogels*

Gelation of a printed hydrogel structure is essential for retaining/preserving its shape. The gelation can either be physical, chemical, or a combination of both processes. These crosslinking methods are used to prepare hydrogels with specific properties, such as mechanical modulus, gelation time and biocompatibility.

(i) Physical crosslinking

In bioprinting, the most prominent hydrogel class is physically crosslinked hydrogels. Physical crosslinking mechanisms are based on entanglements of high molecular polymer chains, hydrogen bridges, ionic interactions, or hydrophobic interactions. Physically crosslinked hydrogels have excellent compatibility with living cells as well as fragile molecules. One of the reasons that physical crosslinking is recommended is that it avoids potentially harmful chemical crosslinking agents. In this subsection, the two dominant physical crosslinking methods are introduced, which are thermo-gelation and self-assembly.

- Thermo-gelation

Hydrophobic interactions can activate thermo-sensitive gelation if temperature changes [60, 61]. The most commonly used thermo-sensitive hydrogels are poly(N-isopropylacrylamide) (PNIPAAm) and poly(ethylene oxide)-poly (propylene oxide)-poly(ethylene oxide) (PEO–PPO–PEO, known as Pluronics). Unfortunately, potential cytotoxicity and non-degradability of these hydrogels restrict their applications in TE. For instance, the viability of HdpG2 cells decreases rapidly when the cells are soaked in Pluronic F127 solution. Cells that were encapsulated into hydrogels with F127 solutions caused cell death only within five days [62]. Additionally, the cytotoxicity is low in thermo-sensitive hydrogels e.g. graft or block copolymers containing hydrophobic polylactic acid (PLA) and hydrophilic PEO moieties [63, 64].

- Self-assembly

A novel concept now adopted in bio-hydrogel designs is supramolecular self-assembly. The coiled-coil pattern is utilised in the design of physically crosslinked hydrogels. In a protein folding process, usually more than two helical winds are combined to form a superhelix [65]. While coils form during protein folding, gelation is triggered [66-68]. The gelation is affected by both the number of coiled-coil grafts and their lengths [69]. The self-assembling properties and thermal stability are related to electrostatic and hydrophobic interactions, which can be controlled by handling the block length of the coiled-coil domains and the amino acid sequences [70].

(ii) Chemical crosslinking

A significant shortcoming of the hydrogels using physical crosslinking is their poor mechanical properties, which may reduce stability of a printed construct. Therefore, chemical crosslinking post-processing has increasingly been used to improve stability and printability of the hydrogels as well as provide high mechanical strength and good handling properties. In general, chemical crosslinking is achieved by mixing gel precursors and two low viscous solutions. Thus, the crosslinking reaction is initiated, leading to a constant increase of viscosity. The representative methods include radical polymerisation and functional groups.

• Crosslinking by radical polymerisation

Free radical polymerisation is a popular method to prepare hydrogels for bioapplications [71, 72]. Hydrogels are formed through polymerisation of vinyl-bearing macromers using redox or thermal initiators, or through photopolymerisation using an ultraviolet (UV) light. The most commonly used natural polymers include chitosan [73], hyaluronic acid [74] and dextran [75], and the synthetic polymers are PVA and PEG [76-78]. Fast crosslinking rates is the major advantage of photo-initiation but cells that are exposed to UV for long times at high intensity may induce disadvantageous effect on cellular metabolic activity [79]. Moreover, the crosslinking process usually releases heat, which can lead to cellular necrosis [80]. Based on this reason, the intensity of the UV light is restricted to approximately 5–10 mW/cm^2 so as to prevent potential cell damage [81]. *In vivo* polymerisation by UV has not been successful owing to very low tissue penetration and absorption rates of the UV light by the skin (> 99%) [81]. Alternative methods are redox- or thermal-initiated polymerisation. Hong *et al.* [82] and Zhu and Ding [83] used the redox-initiator ammonium peroxydisulfate and N,N,N'N'-tetramethyl ethylenediamine (TEMED) to prepare hydrogels. Increasing the concentration of the initiator led to the enhanced mechanical properties and the reduced gelation time. Nevertheless, a high concentration of initiator (10 mM) can also induce high cytotoxicity, and low cell viability (< 30%) was observed after a cell culturing time of 4 days [82]. Therefore, there is a need to develop a more appropriate method for free radical polymerisation for hydrogel preparations.

• Crosslinking by functional groups

Reactions between functional groups in water-soluble macromonomers or monomers can be utilised to prepare hydrogels. Typical reactions include Schiff-based formation, peptide ligation, Michael-type additions and "click" chemistry.

Schiff-based formation between an amino group and an aldehyde is usually used to prepare crosslinked hydrogels [84, 85]. Glutaraldehyde as a crosslinker is frequently used. However, glutaraldehyde is a toxic material even at low concentrations and it can leach out into the human body during matrix degradation, which inhibits cell growth. Therefore, hydrogels prepared by glutaraldehyde crosslinking should be subjected to extensive extraction in order to remove unreacted reagent. Additionally, aldehyde-containing compounds and a nontoxic polymer (e.g. hyaluronic acid) are coupled together for the purpose of avoiding the toxicity related to the use of glutaraldehyde [86]. Moreover, reactive aldehyde groups can be formed by oxidation of a polysaccharide e.g. alginate [87], dextran [84] and hyaluronic acid [85]. As Schiff bases are likely to degrade via hydrolysis of the imine bond even at low pH, adding basic components such as borax will enable Schiff-based formation to generate relatively stable hydrogels with reduced gelation times [87].

Chemical peptide ligation is used for the synthesis of enzymes and proteins, which utilises the chemoselective reaction of two segments of unprotected peptide [88]. For the preparation of hydrogels, peptide ligation is based on the aldehyde groups of NH2-terminal cysteine moieties of peptide dendrons and PEG derivatives to form thiazolidine rings (see Fig. 4.15a) [89]. The reactions take place under mild conditions and gelation occurs within a few minutes. Nevertheless, these hydrogels only stay intact for about one week owing to the reversible thiazolidine ring formation. More stable hydrogels can be obtained by using PEG together with end-capped ester-aldehyde groups as opposed to aldehyde groups via pseudoproline ring formation (Fig. 4.15b) [90]. These hydrogels can retain the shapes and sizes for more than 6 months with less than 10% weight loss.

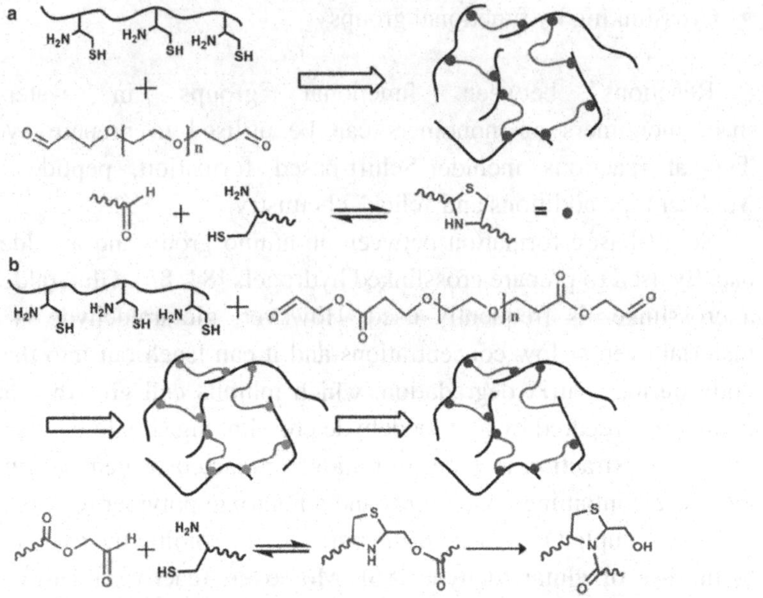

Fig. 4.15. Preparation of hydrogels via peptide ligation [91] (Copyright 2013, with permission of Springer).

The Michael addition reaction between an electrophile (vinyl/acrylate/maleimide group) and a nucleophile (an amine or a thiol group) is another method for hydrogel preparation, particularly for injectable hydrogels (see Fig. 4.16). Hydrogels are formed by mixing two polymers bearing nucleophilic with electrophilic groups. Michael reactions have successfully been used to conjugate many polymers such as dextran [92, 93], PVA [93], hyaluronic acid [94, 95] and PEG [96, 97] with the above groups for hydrogel preparation. Thiol-bearing functional peptides are incorporated to form bio-functional hydrogels, providing enhanced cell adhesion and matrix production [98, 99]. In general, hydrogels produced by this method possess moderate mechanical strength and moderate gelation times (< 0.5 to ~60 min). It is noted that material properties can be adjusted by changing the crosslinking density and the reactivity of the functional groups. Due to Michael addition reactions taking place in mild conditions, the reactions do not significantly affect cell viability during hydrogel formation. Incorporated cells in hydrogels can normally remain viable from days to months [100]. However, some attention has to be paid on the use of an excess of thiol functional groups since thiols may lead to cell death [101].

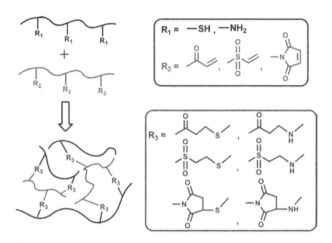

Fig. 4.16. Strategy of preparation of injectable hydrogels via Michael-type additions [91] (Copyright 2010, with permission of Springer).

'Click chemistry' is a highly efficient quantitative reaction, which is conducted at physiological temperatures and pH by the copper-catalysed 1,3-dipolar cycloaddition of alkyne and azide moieties [102]. Click chemistry has been applied to the synthesis of polymer networks from natural polymers like hyaluronic acid [103], and functionalised synthetic polymers such as poly(*N*-isopropylacrylamide-*co*-hydroxylethyl methacrylate) (P(NIPAAm-*co*-HEMA) [104], PVA [105], PEG [106]. Faster gelation is induced by low degree of substitution with active pendant groups. As copper is toxic to most mammalian and bacterial cells [107], it may be dangerous in some cases when performing copper-catalysed click chemistry inside living cells [108]. However, it is difficult to remove copper catalyst from hydrogels. As a result, copper-free click reactions, as a new way of preparing hydrogels for tissue engineering, have gradually gained attention in recent years. Table 4.4 outlines the advantages and disadvantages of crosslinking methods via reactions between functional groups.

Table 4.4. Summary of methods of crosslinking by functional groups.

Chemical reaction	Functional group	Advantage	Disadvantage
Schiff-based formation	Amine/hydrazide & aldehyde	Easy to incorporate and crosslink proteins and amine-bearing peptides	Schiff-based linkage are unstable at low pH Aldehyde may induce side reactions in the body
Peptide ligation	N-terminal cysteine & aldehyde	High efficiency of crosslinking High substrate specificity Mild reaction conditions	Over complicated synthesis procedures of peptides owing to protection and deprotection steps
Michael-type addition	Acrylate/vinyl sulfone & thiol/amine	Tunable properties Appropriate for cell encapsulation Mild reaction conditions	Cell viability may be affected by unreacted thiol groups
Click chemistry	Azide & alkyne	High reaction efficiency	Catalyst contains toxic Cu

(iii)Combining physical and chemical crosslinking

Owing to the reversible interactions, the mechanical properties of hydrogels formed physically in-situ are normally lower than those chemically formed hydrogels. Increasing the molecular weight of polymers as well as the crosslinking density can improve the mechanical properties. However this increases the difficulty in handling accordingly due to the viscosity of hydrogel precursors. Chemically in-situ formed hydrogels usually possess much higher mechanical properties but biologically unfavourable compounds are involved in preparation processes, which can result in bio-incompatible materials. Combining both physical and chemical crosslinking allows materials with improved mechanical and physical properties to be obtained without compromising on biocompatibility. For instance, in the design of in-situ hydrogels by combining photopolymerisation and stereocomplexation, an 8-arm PEG–

PDLLA and an 8-arm PEG–PLLA, stereocomplexed hydrogels and partly functionalised with methacrylate groups (40%) are formed upon mixing (see Fig. 4.17) [81]. UV-irradiation can be used to post-crosslink these hydrogels. These double-crosslinked hydrogels exhibit prolonged degradation times and increased mechanical moduli in comparison to those hydrogels purely formed by stereocomplexation. The initiator concentrations for photopolymerisation (i.e. 0.003 wt%) are much lower than those of conventional photocrosslinking systems (0.05 wt%), which significantly reduces the risks of cell damage arising from the heating effect. Other similar methods are combinations of inclusive complexation, stereocomplexation, thermo-gelation, Michael addition reactions, and photopolymerisation [109-111]. Combining two crosslinking mechanisms in a single hydrogel system provides faster gelation times, improves biocompatibility with cells and proteins and allows better control on hydrogel properties.

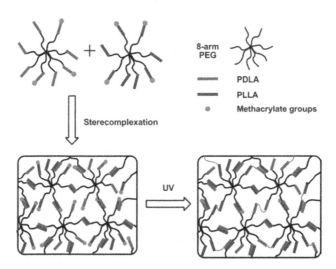

Fig. 4.17. Schematic illustration of hydrogel preparation based on methacrylated PG-PDLA and PEG-PLLA by stereocomplexation and post-UV irradiation [91] (Copyright 2010, with permission of Springer).

4.4.3.3 *Swelling behaviour*

The physical behaviour of hydrogels depends on their dynamic swelling and equilibrium in aqueous solutions and in water. Water molecules in a swollen hydrogel are either bound to hydrophotic/polar polymer chains or exist as free molecules filling up the network of chains. In general, solutes only diffuse within the space taken up by free water molecules, but some chaotropic solutes may destruct the bound water molecules when diffusing through the gel [33]. Swelling contractile characteristics of hydrogels are particularly important factors to be taken into account in wound healing treatments [112]. This is because excessively swelling or contractile materials can potentially lead to slowed healing or fibrotic scarring.

There are two principal opposing forces that behave during a swelling process of hydrogels, which are osmotic pressures and elastic contraction. When using a semi-permeable membrane to separate two solutions with different concentrations, the solvent will tend to diffuse through the membrane from the less concentrated to the more concentrated solution if the membrane is permeable to the smaller solvent molecules but not to the larger solute molecules. This process is named osmosis. Due to the favourable hydrogen bonding between the water and the polymer chains, the water diffuses into the hydrogel network, which leads to osmotic pressures. This results in a pressure built-up within the network. On the other hand, the swelling is also restrained by the elastic contraction of the hydrogel network. The net force resulting from osmotic pressures and elastic contraction is known as the swelling pressure. The following factors determine the swelling properties of a hydrogel: type of polymer (chemical composition, non-ionic, anionic or cationic), hydrophilicity of polymer chains, network structure and crosslinking density, polarity of substituent groups, crystallinity, pH in the aqueous medium, additives (e.g. surfactants and salts) and temperature.

Basically, highly water-swollen hydrogels involve those of cellulose derivatives, PVA, poly(Nvinyl 2-pyrrolidone) (PNVP), and PEG, among others. Moderately and poorly water-swollen hydrogels are those of PHEMA and many of its copolymers. Swelling properties can also be tailored by copolymerising a basic hydrophilic monomer with other more or less hydrophilic monomers. However, it should be noted that the change of swelling properties may influence the solute diffusion coefficient, surface properties and molecule mobility, optical and mechanical properties.

4.4.3.4 *Solute transportation*

Solute transportation determines how cellular products, nutrients and waste are exchanged within a scaffold. Normally, the only driving force in a solution is diffusion as the pores within the gels are too small for convection. In ionic gels, research has discovered that culture condition and hydrogel mesh size are the two primary factors for diffusion. In biological systems, the hydrophilic polymer and their solutes frequently interacts, thus they become important factors determining the transport behaviour. The interactions tend to decrease transport rate of solutes into the hydrogel.

4.5 Integrative Support Materials

Although hydrogels have been widely accepted as biomaterials used in bioprinting due to its biocompatibility, their weak mechanical strength has restricted further applications. A few research groups have incorporated hard thermoplastic materials such as PCL and PLGA to enhance mechanical properties whilst providing a better shape fidelity. As these materials are synthetic, their degradation rates and cytotoxic effects can be well controlled. The use of both hard and soft polymers also enhances cell adhesion and proliferation on scaffolds.

Most hybrid bioprinting systems are comprised of two primary components: (i) dispensing thermoplastic – melt plotting system; and (ii) dispensing hydrogel/cell materials. Table 4.5 summarises some of the current biomaterials used in bioprinting along with the different printing techniques used.

Table 4.5. Popular hydrogels used in bioprinting.

Material	Bioprinting technique	Reference
Natural hydrogel		
Alginate	Inkjet Printing	[113]
	Inkjet Printing	[114]
	Laser Induced Forward Transfer (LIFT) / Biological Laser Printing (BioLP^{TM})	[115]
	Pneumatic Extrusion	[116]
	Pneumatic Extrusion	[117]
	Pneumatic Extrusion	[118]
	Pneumatic Extrusion	[119]
Matrigel	BioLP^{TM}/LIFT	[115]
	Pneumatic Extrusion	[120]
Collagen	BioLP^{TM} / LIFT	[121]
	Inkjet Printing	[122]
Gelatin	BioLP^{TM}/LIFT	[123]
	Positive Displacement Extrusion	[41]
Gel-MA	Dynamic Optical Projection Stereolithography (DOPsL)	[40]
Hyaluronan	Positive Displacement Extrusion	[55]
Synthetic Hydrogel		
PEG	Accoustic Droplet Ejection	[124]
	Stereolithography (SLA)	[125]
	Dynamic Optical Projection Stereolithography (DOPsL)	[126]
Hybrid		
PCL, Alginate	Melt-Plotting System, Pneumatic Extrusion	[117, 127]
PCL, Fibrinogen, Collagen	Electro-spinning Apparatus, Inkjet Printing	[113]
PCL, PLGA, HA, gelatin, collagen	Melt-Plotting System, Pneumatic Extrusion	[128]

References

[1]   A. R. Boccaccini and J. E. Gough, *Tissue engineering using ceramics and polymers*. Cambridge, England: Woodhead Publishing Limited and CRC Press LLC, 2007.

[2]   S. Dumitriu, *Polymeric biomaterials, revised and expanded*. New York, USA: Marcel Dekker, Inc., 2001.

[3]   T. Ogawa, T. Akazawa, and Y. Tabata, "In vitro proliferation and chondrogenic differentiation of rat bone marrow stem cells cultured with gelatin hydrogel microspheres for TGF-β1 release," *Journal of Biomaterials Science, Polymer Edition*, vol. 21, pp. 609-621, 2010.

[4]   A. S. Hoffman, "Hydrogels for biomedical applications," *Advanced drug delivery reviews*, vol. 54, pp. 3-12, 2002.

[5]   M. B. Huglin and D. C. Yip, "An alternative method of determining the water content of hydrogels," *Die Makromolekulare Chemie, Rapid Communications*, vol. 8, pp. 237-242, 1987.

[6]   S. Nagaoka, H. Tanzawa, and J. Suzuki, "Cell proliferation on hydrogels," *In vitro cellular & developmental biology*, vol. 26, pp. 51-56, 1990.

[7]   J. S. Katz and J. A. Burdick, "Light-Responsive Biomaterials: Development and Applications," *Macromolecular bioscience*, vol. 10, pp. 339-348, 2010.

[8]   J. R. Davis, *Handbook of materials for medical devices*. Geauga, Ohio, USA: ASM international, 2003.

[9]   D. Williams, "A model for biocompatibility and its evaluation," *Journal of biomedical engineering*, vol. 11, pp. 185-191, 1989.

[10]   W. Y. Yeong, C. K. Chua, K. F. Leong, and M. Chandrasekaran, "Rapid prototyping in tissue engineering: challenges and potential," *Trends in Biotechnology*, vol. 22, pp. 643-652, 2004.

[11]   H. J. Sung, C. Meredith, C. Johnson, and Z. S. Galis, "The effect of scaffold degradation rate on three-dimensional cell growth and angiogenesis," *Biomaterials*, vol. 25, pp. 5735-5742, 2004.

[12]   L. L. Hench, R. J. Splinter, W. Allen, and T. Greenlee, "Bonding mechanisms at the interface of ceramic prosthetic materials," *Journal of Biomedical Materials Research*, vol. 5, pp. 117-141, 1971.

[13]   L. Bacakova, E. Filova, F. Rypacek, V. Svorcik, and V. Stary, "Cell adhesion on artificial materials for tissue engineering," *Physiol Res*, vol. 53, pp. S35-S45, 2004.

[14]   V. Premnath, W. Harris, M. Jasty, and E. Merrill, "Gamma sterilization of UHMWPE articular implants: an analysis of the oxidation problem," *Biomaterials*, vol. 17, pp. 1741-1753, 1996.

[15]   J. C. Middleton and A. J. Tipton, "Synthetic biodegradable polymers as orthopedic devices," *Biomaterials,* vol. 21, pp. 2335-2346, 2000.

[16]   H. H. Lu, J. A. Cooper Jr, S. Manuel, J. W. Freeman, M. A. Attawia, F. K. Ko, *et al.,* "Anterior cruciate ligament regeneration using braided biodegradable scaffolds: in vitro optimization studies," *Biomaterials,* vol. 26, pp. 4805-4816, 2005.

[17]   J. A. Cooper, H. H. Lu, F. K. Ko, J. W. Freeman, and C. T. Laurencin, "Fiber-based tissue-engineered scaffold for ligament replacement: design considerations and in vitro evaluation," *Biomaterials,* vol. 26, pp. 1523-1532, 2005.

[18]   J. Bergsma, F. Rozema, R. Bos, G. Boering, W. De Bruijn, and A. Pennings, "In vivo degradation and biocompatibility study of in vitro pre-degraded as-polymerized polylactide particles," *Biomaterials,* vol. 16, pp. 267-274, 1995.

[19]   V. Boix, "Polylactic acid implants. A new smile for lipoatrophic faces?," *Aids,* vol. 17, pp. 2533-2535, 2003.

[20]   D. Williams and E. Mort, "Enzyme-accelerated hydrolysis of polyglycolic acid," *Journal of bioengineering,* vol. 1, p. 231, 1977.

[21]   J. F. Nelson, H. G. Stanford, and D. E. Cutright, "Evaluation and comparisons of biodegradable substances as osteogenic agents," *Oral Surgery, Oral Medicine, Oral Pathology,* vol. 43, pp. 836-843, 1977.

[22]   J. O. Hollinger, "Preliminary report on the osteogenic potential of a biodegradable copolymer of polyactide (PLA) and polyglycolide (PGA)," *Journal of biomedical materials research,* vol. 17, pp. 71-82, 1983.

[23]   M. Taylor, A. Daniels, K. Andriano, and J. Heller, "Six bioabsorbable polymers: in vitro acute toxicity of accumulated degradation products," *Journal of Applied Biomaterials,* vol. 5, pp. 151-157, 1994.

[24]   R. L. Kronenthal, Z. Oser, and E. Martin, *Biodegradable polymers in medicine and surgery* vol. 8. New York, US: Springer, 1975.

[25]   P. A. Gunatillake and R. Adhikari, "Biodegradable synthetic polymers for tissue engineering," *Eur Cell Mater,* vol. 5, pp. 1-16, 2003.

[26] P. Gunatillake, R. Mayadunne, and R. Adhikari, "Recent developments in biodegradable synthetic polymers," *Biotechnology annual review*, vol. 12, pp. 301-347, 2006.

[27] R. A. Miller, J. M. Brady, and D. E. Cutright, "Degradation rates of oral resorbable implants (polylactates and polyglycolates): rate modification with changes in PLA/PGA copolymer ratios," *Journal of biomedical materials research*, vol. 11, pp. 711-719, 1977.

[28] L. S. Nair and C. T. Laurencin, "Biodegradable polymers as biomaterials," *Progress in polymer science*, vol. 32, pp. 762-798, 2007.

[29] A. S. Posner, "Crystal chemistry of bone mineral," *Physiological reviews*, vol. 49, pp. 760-792, 1969.

[30] H. Oonishi, L. Hench, J. Wilson, F. Sugihara, E. Tsuji, M. Matsuura, *et al.*, "Quantitative comparison of bone growth behavior in granules of Bioglass®, A-W glass-ceramic, and hydroxyapatite," *Journal of biomedical materials research*, vol. 51, pp. 37-46, 2000.

[31] J. Barralet, T. Gaunt, A. Wright, I. Gibson, and J. Knowles, "Effect of porosity reduction by compaction on compressive strength and microstructure of calcium phosphate cement," *Journal of biomedical materials research*, vol. 63, pp. 1-9, 2002.

[32] H. Yoshikawa, N. Tamai, T. Murase, and A. Myoui, "Interconnected porous hydroxyapatite ceramics for bone tissue engineering," *Journal of The Royal Society Interface*, vol. 6, pp. S341-S348, 2009.

[33] B. D. Ratner, A. S. Hoffman, F. J. Schoen, and J. E. Lemons, *Biomaterials science: an introduction to materials in medicine*, 3rd ed. London, United Kingdom: Elsevier, 2013.

[34] A. Weinstein, J. Klawitter, and S. Cook, "Implant-bone interface characteristics of bioglass dental implants," *Journal of biomedical materials research*, vol. 14, pp. 23-29, 1980.

[35] C. L. Tisdel, V. M. Goldberg, J. A. Parr, J. S. Bensusan, L. S. Staikoff, and S. Stevenson, "The influence of a hydroxyapatite and tricalcium-phosphate coating on bone growth into titanium fiber-metal implants," *The Journal of bone and joint surgery. American volume*, vol. 76, p. 159, 1994.

[36]   C. H. Lee, A. Singla, and Y. Lee, "Biomedical applications of collagen," *International journal of pharmaceutics,* vol. 221, pp. 1-22, 2001.

[37]   C. Lee, A. Grodzinsky, and M. Spector, "The effects of cross-linking of collagen-glycosaminoglycan scaffolds on compressive stiffness, chondrocyte-mediated contraction, proliferation and biosynthesis," *Biomaterials,* vol. 22, pp. 3145-3154, 2001.

[38]   S. N. Park, J. C. Park, H. O. Kim, M. J. Song, and H. Suh, "Characterization of porous collagen/hyaluronic acid scaffold modified by 1-ethyl-3-(3-dimethylaminopropyl) carbodiimide cross-linking," *Biomaterials,* vol. 23, pp. 1205-1212, 2002.

[39]   B. Duan, L. A. Hockaday, K. H. Kang, and J. T. Butcher, "3D bioprinting of heterogeneous aortic valve conduits with alginate/gelatin hydrogels," *Journal of Biomedical Materials Research Part A,* vol. 101, pp. 1255-1264, 2013.

[40]   P. Soman, P. H. Chung, A. P. Zhang, and S. Chen, "Digital microfabrication of user-defined 3D microstructures in cell-laden hydrogels," *Biotechnology and bioengineering,* vol. 110, pp. 3038-3047, 2013.

[41]   J. Visser, B. Peters, T. J. Burger, J. Boomstra, W. J. Dhert, F. P. Melchels, *et al.,* "Biofabrication of multi-material anatomically shaped tissue constructs," *Biofabrication,* vol. 5, p. 035007, 2013.

[42]   T. Billiet, E. Gevaert, T. De Schryver, M. Cornelissen, and P. Dubruel, "The 3D printing of gelatin methacrylamide cell-laden tissue-engineered constructs with high cell viability," *Biomaterials,* vol. 35, pp. 49-62, 2014.

[43]   E. Dare, S. Vascotto, D. Carlsson, M. Hincke, and M. Griffith, "Differentiation of a fibrin gel encapsulated chondrogenic cell line," *The International journal of artificial organs,* vol. 30, p. 619, 2007.

[44]   F. A. Johnson, D. Q. Craig, and A. D. Mercer, "Characterization of the block structure and molecular weight of sodium alginates," *Journal of pharmacy and pharmacology,* vol. 49, pp. 639-643, 1997.

[45]   E. Khor and L. Y. Lim, "Implantable applications of chitin and chitosan," *Biomaterials,* vol. 24, pp. 2339-2349, 2003.

[46]   B. Alberts, D. Bray, J. Lewis, M. Raff, K. Roberts, and J. D. Watson, *Molecular biology of the cell,* 3rd ed. New York, US: Garland Science, 1994.

[47]   B. V. Slaughter, S. S. Khurshid, O. Z. Fisher, A. Khademhosseini, and N. A. Peppas, "Hydrogels in regenerative medicine," *Advanced Materials,* vol. 21, pp. 3307-3329, 2009.

[48]   J. S. Belkas, C. A. Munro, M. S. Shoichet, M. Johnston, and R. Midha, "Long-term in vivo biomechanical properties and biocompatibility of poly (2-hydroxyethyl methacrylate-co-methyl methacrylate) nerve conduits," *Biomaterials,* vol. 26, pp. 1741-1749, 2005.

[49]   R. H. Schmedlen, K. S. Masters, and J. L. West, "Photocrosslinkable polyvinyl alcohol hydrogels that can be modified with cell adhesion peptides for use in tissue engineering," *Biomaterials,* vol. 23, pp. 4325-4332, 2002.

[50]   N. A. Alcantar, E. S. Aydil, and J. N. Israelachvili, "Polyethylene glycol–coated biocompatible surfaces," *Journal of biomedical materials research,* vol. 51, pp. 343-351, 2000.

[51]   H. A. Barnes, J. F. Hutton, and K. Walters, *An introduction to rheology* vol. 3. Oxford, UK: Elsevier, 1989.

[52]   R. Rezende, M. Rezende, R. Maciel Filho, P. Bartolo, and A. Mendes, "Genetic Algorithms for Optimising Alginated-Scaffolds for Tissue Engineering," *CHEMICAL ENGINEERING,* vol. 17, pp. 1305-1309, 2009.

[53]   C. M. Smith, J. J. Christian, W. L. Warren, and S. K. Williams, "Characterizing environmental factors that impact the viability of tissue-engineered constructs fabricated by a direct-write bioassembly tool," *Tissue engineering,* vol. 13, pp. 373-383, 2007.

[54]   X. Cui, K. Breitenkamp, M. Finn, M. Lotz, and D. D. D'Lima, "Direct human cartilage repair using three-dimensional bioprinting technology," *Tissue Engineering Part A,* vol. 18, pp. 1304-1312, 2012.

[55]   A. Skardal, J. Zhang, and G. D. Prestwich, "Bioprinting vessel-like constructs using hyaluronan hydrogels crosslinked with tetrahedral polyethylene glycol tetracrylates," *Biomaterials,* vol. 31, pp. 6173-6181, 2010.

[56]　M. Guvendiren, H. D. Lu, and J. A. Burdick, "Shear-thinning hydrogels for biomedical applications," *Soft Matter,* vol. 8, pp. 260-272, 2012.

[57]　R. A. Rezende, P. J. Bártolo, and A. Mendes, "Rheological behavior of alginate solutions for biomanufacturing," *Journal of applied polymer science,* vol. 113, pp. 3866-3871, 2009.

[58]　J. Tang, M. A. Tung, and Y. Zeng, "Compression strength and deformation of gellan gels formed with mono-and divalent cations," *Carbohydrate Polymers,* vol. 29, pp. 11-16, 1996.

[59]　J. Malda, J. Visser, F. P. Melchels, T. Jüngst, W. E. Hennink, W. J. Dhert, *et al.*, "25th anniversary article: engineering hydrogels for biofabrication," *Advanced Materials,* vol. 25, pp. 5011-5028, 2013.

[60]　W. Hennink and C. Van Nostrum, "Novel crosslinking methods to design hydrogels," *Advanced drug delivery reviews,* vol. 64, pp. 223-236, 2012.

[61]　E. Ruel Gariépy and J. C. Leroux, "In situ-forming hydrogels—review of temperature-sensitive systems," *European Journal of Pharmaceutics and Biopharmaceutics,* vol. 58, pp. 409-426, 2004.

[62]　S. F. Khattak, S. R. Bhatia, and S. C. Roberts, "Pluronic F127 as a cell encapsulation material: utilization of membrane-stabilizing agents," *Tissue engineering,* vol. 11, pp. 974-983, 2005.

[63]　B. Jeong, Y. H. Bae, D. S. Lee, and S. W. Kim, "Biodegradable block copolymers as injectable drug-delivery systems," *Nature,* vol. 388, pp. 860-862, 1997.

[64]　B. Jeong, K. M. Lee, A. Gutowska, and Y. H. An, "% Thermogelling Biodegradable Copolymer Aqueous Solutions for Injectable Protein Delivery and Tissue Engineering," *Biomacromolecules,* vol. 3, pp. 865-868, 2002.

[65]　Y. B. Yu, "Coiled-coils: stability, specificity, and drug delivery potential," *Advanced drug delivery reviews,* vol. 54, pp. 1113-1129, 2002.

[66]　J. Yang, C. Xu, C. Wang, and J. Kopecek, "Refolding hydrogels self-assembled from N-(2-hydroxypropyl) methacrylamide graft copolymers by antiparallel coiled-coil formation," *Biomacromolecules,* vol. 7, pp. 1187-1195, 2006.

[67]　J. Yang, C. Xu, P. Kopečková, and J. Kopeček, "Hybrid Hydrogels Self-Assembled from HPMA Copolymers Containing

Peptide Grafts," *Macromolecular bioscience,* vol. 6, pp. 201-209, 2006.

[68] C. Wang, J. Kopecek, and R. J. Stewart, "Hybrid hydrogels cross-linked by genetically engineered coiled-coil block proteins," *Biomacromolecules,* vol. 2, pp. 912-920, 2001.

[69] C. Wang, R. J. Stewart, and J. Kopeček, "Hybrid hydrogels assembled from synthetic polymers and coiled-coil protein domains," *Nature,* vol. 397, pp. 417-420, 1999.

[70] C. Xu, V. Breedveld, and J. Kopecek, "Reversible hydrogels from self-assembling genetically engineered protein block copolymers," *Biomacromolecules,* vol. 6, pp. 1739-1749, 2005.

[71] J. L. Ifkovits and J. A. Burdick, "Review: photopolymerizable and degradable biomaterials for tissue engineering applications," *Tissue engineering,* vol. 13, pp. 2369-2385, 2007.

[72] K. T. Nguyen and J. L. West, "Photopolymerizable hydrogels for tissue engineering applications," *Biomaterials,* vol. 23, pp. 4307-4314, 2002.

[73] Y. Hong, Z. Mao, H. Wang, C. Gao, and J. Shen, "Covalently crosslinked chitosan hydrogel formed at neutral pH and body temperature," *Journal of Biomedical Materials Research Part A,* vol. 79, pp. 913-922, 2006.

[74] Y. D. Park, N. Tirelli, and J. A. Hubbell, "Photopolymerized hyaluronic acid-based hydrogels and interpenetrating networks," *Biomaterials,* vol. 24, pp. 893-900, 2003.

[75] S. H. Kim, C. Y. Won, and C. C. Chu, "Synthesis and characterization of dextran-maleic acid based hydrogel," *Journal of biomedical materials research,* vol. 46, pp. 160-170, 1999.

[76] S. J. Bryant and K. S. Anseth, "Hydrogel properties influence ECM production by chondrocytes photoencapsulated in poly (ethylene glycol) hydrogels," *Journal of Biomedical Materials Research,* vol. 59, pp. 63-72, 2002.

[77] P. J. Martens, S. J. Bryant, and K. S. Anseth, "Tailoring the degradation of hydrogels formed from multivinyl poly (ethylene glycol) and poly (vinyl alcohol) macromers for cartilage tissue engineering," *Biomacromolecules,* vol. 4, pp. 283-292, 2003.

[78] D. Mawad, P. J. Martens, R. A. Odell, and L. A. Poole-Warren, "The effect of redox polymerisation on degradation and cell responses to poly (vinyl alcohol) hydrogels," *Biomaterials,* vol. 28, pp. 947-955, 2007.

[79] S. J. Bryant, C. R. Nuttelman, and K. S. Anseth, "Cytocompatibility of UV and visible light photoinitiating systems on cultured NIH/3T3 fibroblasts in vitro," *Journal of Biomaterials Science, Polymer Edition,* vol. 11, pp. 439-457, 2000.

[80] J. Łukaszczyk, M. Śmiga, K. Jaszcz, H. J. P. Adler, E. Jähne, and M. Kaczmarek, "Evaluation of oligo (ethylene glycol) dimethacrylates effects on the properties of new biodegradable bone cement compositions," *Macromolecular bioscience,* vol. 5, pp. 64-69, 2005.

[81] C. Hiemstra, W. Zhou, Z. Zhong, M. Wouters, and J. Feijen, "Rapidly in situ forming biodegradable robust hydrogels by combining stereocomplexation and photopolymerization," *Journal of the American Chemical Society,* vol. 129, pp. 9918-9926, 2007.

[82] Y. Hong, H. Song, Y. Gong, Z. Mao, C. Gao, and J. Shen, "Covalently crosslinked chitosan hydrogel: Properties of in vitro degradation and chondrocyte encapsulation," *Acta biomaterialia,* vol. 3, pp. 23-31, 2007.

[83] W. Zhu and J. Ding, "Synthesis and characterization of a redox-initiated, injectable, biodegradable hydrogel," *Journal of applied polymer science,* vol. 99, pp. 2375-2383, 2006.

[84] J. Maia, L. Ferreira, R. Carvalho, M. A. Ramos, and M. H. Gil, "Synthesis and characterization of new injectable and degradable dextran-based hydrogels," *Polymer,* vol. 46, pp. 9604-9614, 2005.

[85] K. Y. Lee, E. Alsberg, and D. J. Mooney, "Degradable and injectable poly (aldehyde guluronate) hydrogels for bone tissue engineering," *Journal of biomedical materials research,* vol. 56, pp. 228-233, 2001.

[86] P. Bulpitt and D. Aeschlimann, "New strategy for chemical modification of hyaluronic acid: preparation of functionalized derivatives and their use in the formation of novel biocompatible hydrogels," *Journal of biomedical materials research,* vol. 47, pp. 152-169, 1999.

[87] B. Balakrishnan and A. Jayakrishnan, "Self-cross-linking biopolymers as injectable in situ forming biodegradable scaffolds," *Biomaterials,* vol. 26, pp. 3941-3951, 2005.

[88] G. J. Cotton and T. W. Muir, "Peptide ligation and its application to protein engineering," *Chemistry & biology,* vol. 6, pp. R247-R256, 1999.

[89] M. Wathier, P. J. Jung, M. A. Carnahan, T. Kim, and M. W. Grinstaff, "Dendritic macromers as in situ polymerizing biomaterials for securing cataract incisions," *Journal of the American Chemical Society,* vol. 126, pp. 12744-12745, 2004.

[90] M. Wathier, C. S. Johnson, T. Kim, and M. W. Grinstaff, "Hydrogels formed by multiple peptide ligation reactions to fasten corneal transplants," *Bioconjugate chemistry,* vol. 17, pp. 873-876, 2006.

[91] R. Jin and P. J. Dijkstra, "Hydrogels for tissue engineering applications," in *Biomedical applications of hydrogels handbook,* ed New York, USA: Springer, 2010, pp. 203-225.

[92] C. Hiemstra, L. J. van der Aa, Z. Zhong, P. J. Dijkstra, and J. Feijen, "Novel in situ forming, degradable dextran hydrogels by Michael addition chemistry: synthesis, rheology, and degradation," *Macromolecules,* vol. 40, pp. 1165-1173, 2007.

[93] C. Hiemstra, L. J. van der Aa, Z. Zhong, P. J. Dijkstra, and J. Feijen, "Rapidly in situ-forming degradable hydrogels from dextran thiols through Michael addition," *Biomacromolecules,* vol. 8, pp. 1548-1556, 2007.

[94] C. M. Riley, P. W. Fuegy, M. A. Firpo, X. Zheng Shu, G. D. Prestwich, and R. A. Peattie, "Stimulation of in vivo angiogenesis using dual growth factor-loaded crosslinked glycosaminoglycan hydrogels," *Biomaterials,* vol. 27, pp. 5935-5943, 2006.

[95] X. Zheng Shu, Y. Liu, F. S. Palumbo, Y. Luo, and G. D. Prestwich, "In situ crosslinkable hyaluronan hydrogels for tissue engineering," *Biomaterials,* vol. 25, pp. 1339-1348, 2004.

[96] J. L. Vanderhooft, B. K. Mann, and G. D. Prestwich, "Synthesis and characterization of novel thiol-reactive poly (ethylene glycol) cross-linkers for extracellular-matrix-mimetic biomaterials," *Biomacromolecules,* vol. 8, pp. 2883-2889, 2007.

[97] A. Metters and J. Hubbell, "Network formation and degradation behavior of hydrogels formed by Michael-type addition reactions," *Biomacromolecules,* vol. 6, pp. 290-301, 2005.

[98] M. Lutolf, J. Lauer-Fields, H. Schmoekel, A. Metters, F. Weber, G. Fields, *et al.,* "Synthetic matrix metalloproteinase-sensitive

hydrogels for the conduction of tissue regeneration: engineering cell-invasion characteristics," *Proceedings of the National Academy of Sciences,* vol. 100, pp. 5413-5418, 2003.

[99]   M. P. Lutolf, G. P. Raeber, A. H. Zisch, N. Tirelli, and J. A. Hubbell, "Cell-Responsive Synthetic Hydrogels," *Advanced Materials,* vol. 15, pp. 888-892, 2003.

[100]   J. Kim, I. S. Kim, T. H. Cho, K. B. Lee, S. J. Hwang, G. Tae, *et al.,* "Bone regeneration using hyaluronic acid-based hydrogel with bone morphogenic protein-2 and human mesenchymal stem cells," *Biomaterials,* vol. 28, pp. 1830-1837, 2007.

[101]   C. N. Salinas, B. B. Cole, A. M. Kasko, and K. S. Anseth, "Chondrogenic differentiation potential of human mesenchymal stem cells photoencapsulated within poly (ethylene glycol)-arginine-glycine-aspartic acid-serine thiol-methacrylate mixed-mode networks," *Tissue engineering,* vol. 13, pp. 1025-1034, 2007.

[102]   V. V. Rostovtsev, L. G. Green, V. V. Fokin, and K. B. Sharpless, "A stepwise huisgen cycloaddition process: copper (I)-catalyzed regioselective "ligation" of azides and terminal alkynes," *Angewandte Chemie,* vol. 114, pp. 2708-2711, 2002.

[103]   V. Crescenzi, L. Cornelio, C. Di Meo, S. Nardecchia, and R. Lamanna, "Novel hydrogels via click chemistry: synthesis and potential biomedical applications," *Biomacromolecules,* vol. 8, pp. 1844-1850, 2007.

[104]   X. D. Xu, C. S. Chen, Z. C. Wang, G. R. Wang, S. X. Cheng, X. Z. Zhang, *et al.,* ""Click" chemistry for in situ formation of thermoresponsive P (NIPAAm-co-HEMA)-based hydrogels," *Journal of Polymer Science Part A: Polymer Chemistry,* vol. 46, pp. 5263-5277, 2008.

[105]   D. A. Ossipov and J. Hilborn, "Poly (vinyl alcohol)-based hydrogels formed by "click chemistry"," *Macromolecules,* vol. 39, pp. 1709-1718, 2006.

[106]   M. Malkoch, R. Vestberg, N. Gupta, L. Mespouille, P. Dubois, A. F. Mason, *et al.,* "Synthesis of well-defined hydrogel networks using click chemistry," *Chemical Communications,* pp. 2774-2776, 2006.

[107]   J. A. Johnson, J. M. Baskin, C. R. Bertozzi, J. T. Koberstein, and N. J. Turro, "Copper-free click chemistry for the in situ

crosslinking of photodegradable star polymers," *Chem. Commun.,* pp. 3064-3066, 2008.

[108] A. E. Speers and B. F. Cravatt, "Profiling enzyme activities in vivo using click chemistry methods," *Chemistry & biology,* vol. 11, pp. 535-546, 2004.

[109] G. Niu, H. Zhang, L. Song, X. Cui, H. Cao, Y. Zheng, *et al.,* "Thiol/acrylate-modified PEO-PPO-PEO triblocks used as reactive and thermosensitive copolymers," *Biomacromolecules,* vol. 9, pp. 2621-2628, 2008.

[110] S. A. Robb, B. H. Lee, R. McLemore, and B. L. Vernon, "Simultaneously physically and chemically gelling polymer system utilizing a poly (NIPAAm-co-cysteamine)-based copolymer," *Biomacromolecules,* vol. 8, pp. 2294-2300, 2007.

[111] S. Chatrchyan, G. Hmayakyan, V. Khachatryan, A. Sirunyan, W. Adam, T. Bauer, *et al.,* "The CMS experiment at the CERN LHC," *Journal of Instrumentation,* vol. 3, p. S08004, 2008.

[112] S. V. Murphy, A. Skardal, and A. Atala, "Evaluation of hydrogels for bio-printing applications," *Journal of Biomedical Materials Research Part A,* vol. 101, pp. 272-284, 2013.

[113] T. Xu, K. W. Binder, M. Z. Albanna, D. Dice, W. Zhao, J. J. Yoo, *et al.,* "Hybrid printing of mechanically and biologically improved constructs for cartilage tissue engineering applications," *Biofabrication,* vol. 5, p. 015001, 2013.

[114] C. Xu, W. Chai, Y. Huang, and R. R. Markwald, "Scaffold-free inkjet printing of three-dimensional zigzag cellular tubes," *Biotechnology and bioengineering,* vol. 109, pp. 3152-3160, 2012.

[115] B. Guillotin, A. Souquet, S. Catros, M. Duocastella, B. Pippenger, S. Bellance, *et al.,* "Laser assisted bioprinting of engineered tissue with high cell density and microscale organization," *Biomaterials,* vol. 31, pp. 7250-7256, 2010.

[116] S. Ahn, H. Lee, and G. Kim, "Functional cell-laden alginate scaffolds consisting of core/shell struts for tissue regeneration," *Carbohydrate polymers,* vol. 98, pp. 936-942, 2013.

[117] H. Lee, S. Ahn, L. J. Bonassar, W. Chun, and G. Kim, "Cell-laden poly (ε-caprolactone)/alginate hybrid scaffolds fabricated by an aerosol cross-linking process for obtaining homogeneous cell distribution: fabrication, seeding efficiency, and cell

proliferation and distribution," *Tissue Engineering Part C: Methods,* vol. 19, pp. 784-793, 2013.

[118] N. E. Fedorovich, W. Schuurman, H. M. Wijnberg, H.-J. Prins, P. R. Van Weeren, J. Malda, *et al.,* "Biofabrication of osteochondral tissue equivalents by printing topologically defined, cell-laden hydrogel scaffolds," *Tissue Engineering Part C: Methods,* vol. 18, pp. 33-44, 2012.

[119] J. H. Shim, J. S. Lee, J. Y. Kim, and D. W. Cho, "Bioprinting of a mechanically enhanced three-dimensional dual cell-laden construct for osteochondral tissue engineering using a multi-head tissue/organ building system," *Journal of Micromechanics and Microengineering,* vol. 22, p. 085014, 2012.

[120] J. Snyder, Q. Hamid, C. Wang, R. Chang, K. Emami, H. Wu, *et al.,* "Bioprinting cell-laden matrigel for radioprotection study of liver by pro-drug conversion in a dual-tissue microfluidic chip," *Biofabrication,* vol. 3, p. 034112, 2011.

[121] S. Michael, H. Sorg, C. T. Peck, L. Koch, A. Deiwick, B. Chichkov, *et al.,* "Tissue engineered skin substitutes created by laser-assisted bioprinting form skin-like structures in the dorsal skin fold chamber in mice," *PloS one,* vol. 8, p. e57741, 2013.

[122] T. Xu, W. Zhao, J. M. Zhu, M. Z. Albanna, J. J. Yoo, and A. Atala, "Complex heterogeneous tissue constructs containing multiple cell types prepared by inkjet printing technology," *Biomaterials,* vol. 34, pp. 130-139, 2013.

[123] N. R. Schiele, D. B. Chrisey, and D. T. Corr, "Gelatin-based laser direct-write technique for the precise spatial patterning of cells," *Tissue Engineering Part C: Methods,* vol. 17, pp. 289-298, 2010.

[124] Y. Fang, J. P. Frampton, S. Raghavan, R. Sabahi-Kaviani, G. Luker, C. X. Deng, *et al.,* "Rapid generation of multiplexed cell cocultures using acoustic droplet ejection followed by aqueous two-phase exclusion patterning," *Tissue Engineering Part C: Methods,* vol. 18, pp. 647-657, 2012.

[125] V. Chan, P. Zorlutuna, J. H. Jeong, H. Kong, and R. Bashir, "Three-dimensional photopatterning of hydrogels using stereolithography for long-term cell encapsulation," *Lab on a Chip,* vol. 10, pp. 2062-2070, 2010.

[126] Y. Lu, G. Mapili, G. Suhali, S. Chen, and K. Roy, "A digital micro-mirror device-based system for the microfabrication of

complex, spatially patterned tissue engineering scaffolds," *Journal of Biomedical Materials Research Part A,* vol. 77, pp. 396-405, 2006.

[127] W. Schuurman, V. Khristov, M. Pot, P. Van Weeren, W. Dhert, and J. Malda, "Bioprinting of hybrid tissue constructs with tailorable mechanical properties," *Biofabrication,* vol. 3, p. 021001, 2011.

[128] J.-H. Shim, J. Y. Kim, M. Park, J. Park, and D.-W. Cho, "Development of a hybrid scaffold with synthetic biomaterials and hydrogel using solid freeform fabrication technology," *Biofabrication,* vol. 3, p. 034102, 2011.

Problems

1. What are the key requirements for biomaterials in tissue engineering?
2. Draw the molecular structures of polyglycolic acid (PGA) and polycaprolactone (PCL) and based on the structures, explain why they have different properties, e.g. hydrophilicity, degradation rate and etc.
3. How is hydroxyapatite coated on an orthopaedic implant and why is it needed?
4. Give three examples of polysaccharide-based hydrogels.
5. List the key properties of bio-inks for bioprinting.

Chapter 5

Cell Sources for Bioprinting

Cells are an essential part in the approach of tissue engineering (TE). In the scaffold-based approach, cells are seeded secondarily after the scaffold is constructed. In bioprinting, cells are added onto the construct during the printing process. In both approaches, the selection and supply of cells will affect the success of the implementation. Cells can be categorised in different ways such as source, potential and maturation state. For bioprinting, the added variance comes in the form of the cell format, in which individual or aggregates of cells are added into hydrogel matrix or cells can be deposited as cell spheroids. In this chapter, cell sources, the potential for expansion and differentiation, together with different formats of cells for printing are presented and discussed.

5.1 Cell Sources

5.1.1 *Autologous cells*

Autologous cells are cells (such as a biopsy of skin) that are isolated from the patient and then provided for the patient himself/herself after *in vitro* expansion. For example, in blood marrow transplantation, autologous means the situation in which the donor and recipient is the same person. Patients scheduled for non-emergency surgery may be autologous donors by donating blood for themselves that will be stored and used later after the surgery. The major advantages of using autologous cells are there will be no rejection and disease transmission. The disadvantage of autologous transplantations is that they are highly restricted by limited donor sites.

5.1.2 *Allogeneic cells*

Allogeneic cells are cells that are taken from different individuals of the same species. Two or more individuals are said to be allogeneic to one another when the genes at one or more loci are not identical. Allogeneic cells can be used clinically if the immune rejection is properly managed, such as allogeneic bone marrow transplantations. However, in general, allogeneic cells are not recommended for tissue engineering, unless the detrimental genes can be muted through genetic manipulations.

5.1.3 *Xenogeneic cells*

Xenogeneic denotes individuals or tissues from individuals of different species and hence of disparate cell type. Similar to allogeneic cells, it is difficult to overcome the immune rejection when these cells are used.

5.1.3.1 *Advantages of allogeneic and xenogeneic cells*

- Cell sources can be expanded to include cadavers when using allogeneic cells.
- Xenogeneic cells provide a broad range of cell sources with immediate availability.

5.1.3.2 *Disadvantages of allogeneic and xenogeneic cells*

- The human use of allogeneic and xenogeneic cells is related to some obstacles and the major one is the control of the risks of transmitting known and unknown pathogens.
- Potential risks may exist, such as the possibility of introducing infectious diseases into general population by adaptation in a host which is immune-suppressed.
- The risk of immunological rejection of animal cells has not been completely solved.
- It is difficult to maintain the survival and functions of the xenogeneic cells and tissues in the long term.

• Using immunosuppressive treatment is highly risky owing to the weakening of host defence mechanisms.

5.2 Potential for Expansion and Differentiation

Cells can be divided into different categories based on their potential for expansion and differentiation. In general, two major sources of cells can be used for tissue engineering, which are specialised and stem cells. The potential for differentiation in specialised cells are low even though their availability is high through biopsy. On the other hand, embryonic stem cells being the most potent have the highest expansion capacity while the availability is limited. Figure 5.1 depicts the relationship of expansion capacity and availability between different levels of cells, which illustrates that the level of potency decreases as cells progress in later stages of development.

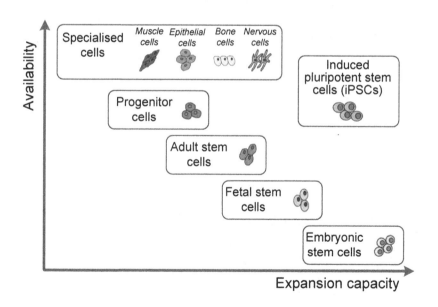

Fig. 5.1. Schematic diagram showing the expansion capacity versus availability between different levels of cells.

5.2.1 *Specialised cells*

Specialised cells are terminally differentiated cells exerting a specific defined physiological function. The phenotype of specialised cells differs from one tissue to another [1]. In general, specialised cells are acquired by dissociation from a tissue donated by others. Autologous, allogeneic or heterologous cells can be used as the donor tissue source [2]. Among these cells, autologous cells are preferable for bioprinting since they are histocompatible and the host rejection can be avoided. As a result, the risks of detrimental side effects induced by immunosuppressive medications can be significantly reduced after the printed constructs are implanted. However, it should be noted that an initial inflammatory response might be caused when using autologous cells.

The embryonic-chick spinal-cord cells are the first specialised cells used in bioprinting, which were deposited by a laser guidance direct write system [3, 4]. Nowadays, more than 30 different types of specialised cells are widely used in bioprinting techniques. These cells are isolated from a number of types of tissue including cardiac [5], nervous [4], epithelial [6, 7], liver [8, 9] and bone [10, 11]. A number of bioprinting techniques have been proved to be viable for printing these cells, including inkjet printing [12], laser-based techniques (i.e. laser-induced forward transfer [13] and laser guidance direct write [4]), extrusion-based techniques [14, 15] and electrostatic-based jetting [16, 17].

5.2.1.1 *Advantages*

- Phenotypically relevant and physiologically functional.
- Straight forward: isolate and use.

5.2.1.2 *Disadvantages*

- Poor *in vitro* differentiation.
- Limited expansion capacity.

5.2.2 *Stem cells*

Stem cells are a class of cells that are capable of both self-renewing and generating differentiated progeny. It is noted that potency for proliferation and differentiation reduces from stem cells to specialised cells. Various types of stem cells can be isolated from different tissues, differentiated and/or expanded *in vitro*. Typically, stem cells include (also see Fig. 5.1) the following:

- Embryonic stem cells (hESCs) derived from blastocysts;
- Somatic or adult stem cells including
  - Mesenchymal stromal/stem cells (MSCs);
  - Haematopoietic progenitor/stem cells (HSCs);
  - Progenitor cells with limited differentiation capacity only responsible for renewal and turnover of normal tissue, such as neurons, skin and lung.
- Induced pluripotent stem cells (iPSCs).

5.2.2.1 *Characteristics of stem cells*

- Embryonic stem cells

Embryonic stem cells are pluripotent and their unique characteristic is they can differentiate to almost every type of cell in a human body. A set of markers for cell surface or marker genes for pluripotency can be used to characterise hESCs. They can be differentiated *in vitro* by genetic modification or using external factors in the culture medium. Nevertheless, *in vitro* differentiation usually results in cell populations with varying levels of heterogeneity. There are, however, ethical debates surrounding the use of human embryonic stem cells.

- Mesenchymal, stromal/stem cells

MSCs are mostly derived from adipose tissue or stroma of bone marrow. Moreover, they can be isolated from a great number of tissue types, such as placenta, liver, blood, tendons and retina. MSCs can differentiate to primarily osteogenic, chondrogenic and adipogenic cell lineages, and mesenchymal lineages, and hence they are identified as lineage-committed cells.

• Tissue specific stem cells

This type of cell has a restricted differentiation capacity. These stem cells usually produce a single or several types of cell highly specific to the particular tissue (e.g. myocytes and astrocytes).

• Haematopoietic stem cells

These cells can be categorised as a special group of tissue-specific stem cells. HSCs can differentiate towards all haematopoietic lineages, lymphoid and myeloid. They widely exist in cord blood in the places where concentration levels are comparable to those of adult bone marrow. In addition, HSCs are largely localised in the red bone marrow and circulate in peripheral blood at a relatively low frequency. They are also found to flow in other tissues such as liver, muscle and spleen at low frequency. On the other hand, they have particularly high mobility to blood compartment, after special treatments with growth factors and chemotherapy.

• Induced pluripotent stem cells

IPSCs are artificially generated cells. Somatic adult cells e.g. skin fibroblasts are reprogrammed to form pluripotent stem cells. In fact, iPSCs and hESCs share many common features. They are pluripotent and possess self-renewing capacity. IPSCs can now be manufactured from various adult cell types and the differentiation capacity is related to the cell type and age.

5.3 Processing of Cells for Bioprinting

In bioprinting, a large number of cells are needed to fabricate or print a biological construct. Different cell bioprinting approaches have been devised. Cells can be printed in the form of suspension or encapsulated within hydrogel to be printed. In the effort to increase the cell density during printing, cells spheroids are fabricated and then directly deposited as a basic unit in a printing process. Figure 5.2 shows the variations of cell format for bioprinting.

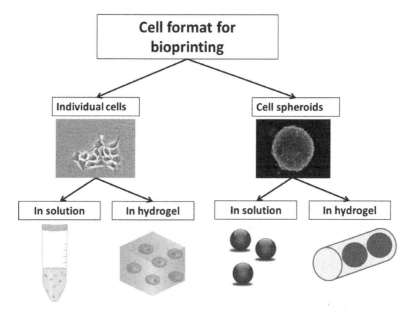

Fig. 5.2. Cell formats for bioprinting includes individual cells or cell spheroids. Different format of cells can be printed in solution or encapsulated within hydrogel.

5.3.1 *Individual cells for bioprinting*

Various suspensions of cells are introduced as follows:

(I) Cells in hydrogels or precursors

- Cartilage progenitor cells are extruded into sodium alginate solution, forming droplets of cell suspension which are later crosslinked using addition of calcium chloride after printing [18].

- Cells are suspended in the mixture of GelMA, photoinitiators and growth medium. The light is selectively projected into the vat containing the above suspension, forming the cell encapsulated hydrogel [19].

- Pre-gel made of decellularised ECM containing human adipose derived stem cells (hASCs) or human inferior turbinate-tissue derived mesenchymal stromal cells (hTMSCs) are printed using an extrusion-based system [20].

(II) Cells in medium

- Bioink containing cells (human embryonic kidney cells and stem cells) in media is deposited in the arrays using valve-based bioprinting technology [21].

- A layer of cell ink containing multipotent mouse bone marrow stromal precursor in the cell medium is spread onto the LAB cartridge. Droplets of cell suspension are ejected onto a biopaper that is located under the cell ink layer [22].

5.3.2 Cell spheroids for bioprinting

5.3.2.1 Cell spheroids

Cell spheroids have been used as an *in vitro* 3D model system in the biomedical and tumour research for several decades. Two primary requirements have to be fulfilled in order to extend the applications of cell spheroids for bioprinting:

- Cell spheroids should be of sufficient quantity and cell density.

- The characteristics of cell spheroids should be well-characterised to be compatible with the bioprinting strategies.

5.3.2.2 *Characteristics of cell spheroids*

Cell spheroids are visco-elastic-plastic soft matter or complex fluid. It is a fundamental principle of solid biodegradable scaffold-free directed tissue self-assembly. Fluidity of cell spheroids is displayed through the following phenomenon (see Fig. 5.3):

- *Rounding*. Spheroids are rounded.

- *Fusion*. Once two aggregates or spheroids are in contact, the interacting cell lines gradually fuse together to generate a larger aggregate with a spherical shape [23]. It has been reported that the fusion kinetics of two tissue explants, which are loaded in a hanging drop, completely fits to fusion kinetics described for two droplets of fluids [24].

- *Enveloping*. This phenomenon takes place when tissue spheroids with different cohesion levels fuse. The lower cohesive spheroid (red) envelopes more cohesive spheroid (green) [24].

- *Sorting*. When a single cell spheroid contains different cell populations, they spontaneously sort out without human intervention. This phenomenon is driven by the different surfaces and interfacial tensions in different cell types [23].

- *Spreading*. Spheroid spreading on a wettable substrate.

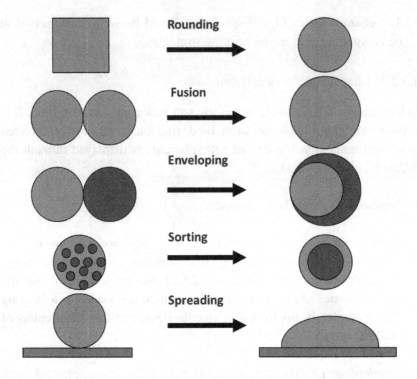

Fig. 5.3. Evidence of fluidity of tissue spheroids.

5.3.2.3 *Preparation for Printing spheroids*

Cells are first obtained from cell cultured after trypsinisation. Upon centrifugation, cell pellets are transferred into cylindrical pipette of approximately 300 to 500 µm. The cylindrical pipette containing cells is incubated overnight. Thereafter, the cylindrical rods of cells are removed from the pipette to obtain multicellular spheroid by cutting up the rod into small clusters. These clusters are gathered and cultured in a gyratory shaker overnight before loading into cylindrical pipette for printing [25]. Alternatively, cylindrical rods of cells can be printed from the pipette.

References

[1]   J. M. Polak and A. E. Bishop, "Stem cells and tissue engineering: past, present, and future," *Annals of the New York Academy of Sciences,* vol. 1068, pp. 352-366, 2006.
[2]   A. Atala, "Tissue engineering and regenerative medicine: concepts for clinical application," *Rejuvenation research,* vol. 7, pp. 15-31, 2004.
[3]   D. J. Odde and M. J. Renn, "Laser-guided direct writing for applications in biotechnology," *Trends in biotechnology,* vol. 17, pp. 385-389, 1999.
[4]   D. J. Odde and M. J. Renn, "Laser-guided direct writing of living cells," *Biotechnology and bioengineering,* vol. 67, pp. 312-318, 2000.
[5]   J. A. Barron, B. R. Ringeisen, H. Kim, B. J. Spargo, and D. B. Chrisey, "Application of laser printing to mammalian cells," *Thin Solid Films,* vol. 453, pp. 383-387, 2004.
[6]   M. Nakamura, A. Kobayashi, F. Takagi, A. Watanabe, Y. Hiruma, K. Ohuchi, *et al.,* "Biocompatible inkjet printing technique for designed seeding of individual living cells," *Tissue engineering,* vol. 11, pp. 1658-1666, 2005.
[7]   T. Boland, X. Tao, B. J. Damon, B. Manley, P. Kesari, S. Jalota, *et al.,* "Drop-on-demand printing of cells and materials for designer tissue constructs," *Materials Science and Engineering: C,* vol. 27, pp. 372-376, 2007.
[8]   R. Saunders, L. Bosworth, J. Gough, B. Derby, and N. Reis, "Selective cell delivery for 3D tissue culture and engineering," *Eur Cells Mater,* vol. 7, pp. 84-5, 2004.
[9]   Y. Nahmias and D. J. Odde, "Micropatterning of living cells by laser-guided direct writing: application to fabrication of hepatic–endothelial sinusoid-like structures," *Nature protocols,* vol. 1, pp. 2288-2296, 2006.
[10]   P. De Coppi, G. Bartsch, M. M. Siddiqui, T. Xu, C. C. Santos, L. Perin, *et al.,* "Isolation of amniotic stem cell lines with potential for therapy," *Nature biotechnology,* vol. 25, pp. 100-106, 2007.
[11]   R. Saunders, J. Gough, and B. Derby, "Ink jet printing of mammalian primary cells for tissue engineering applications," DTIC Document2005.
[12]   T. Xu, J. Jin, C. Gregory, J. J. Hickman, and T. Boland, "Inkjet printing of viable mammalian cells," *Biomaterials,* vol. 26, pp. 93-99, 2005.

[13]   J. Barron, B. Spargo, and B. Ringeisen, "Biological laser printing of three dimensional cellular structures," *Applied Physics A,* vol. 79, pp. 1027-1030, 2004.

[14]   Y. Yan, X. Wang, Y. Pan, H. Liu, J. Cheng, Z. Xiong, *et al.,* "Fabrication of viable tissue-engineered constructs with 3D cell-assembly technique," *Biomaterials,* vol. 26, pp. 5864-5871, 2005.

[15]   D. L. Cohen, E. Malone, H. Lipson, and L. J. Bonassar, "Direct freeform fabrication of seeded hydrogels in arbitrary geometries," *Tissue engineering,* vol. 12, pp. 1325-1335, 2006.

[16]   S. N. Jayasinghe, P. A. Eagles, and A. N. Qureshi, "Electric field driven jetting: an emerging approach for processing living cells," *Biotechnology journal,* vol. 1, pp. 86-94, 2006.

[17]   P. Chandrasekaran, C. Seagle, L. Rice, J. Macdonald, and D. A. Gerber, "Functional analysis of encapsulated hepatic progenitor cells," *Tissue engineering,* vol. 12, pp. 2001-2008, 2006.

[18]   I. T. Ozbolat, H. Chen, and Y. Yu, "Development of 'Multi-arm Bioprinter'for hybrid biofabrication of tissue engineering constructs," *Robotics and Computer-Integrated Manufacturing,* vol. 30, pp. 295-304, 2014.

[19]   P. Soman, P. H. Chung, A. P. Zhang, and S. Chen, "Digital microfabrication of user-defined 3D microstructures in cell-laden hydrogels," *Biotechnology and bioengineering,* vol. 110, pp. 3038-3047, 2013.

[20]   F. Pati, J. Jang, D. H. Ha, S. W. Kim, J. W. Rhie, J. H. Shim, *et al.,* "Printing three-dimensional tissue analogues with decellularized extracellular matrix bioink," *Nature Communications,* vol. 5, 2014.

[21]   A. Faulkner Jones, S. Greenhough, J. A. King, J. Gardner, A. Courtney, and W. Shu, "Development of a valve-based cell printer for the formation of human embryonic stem cell spheroid aggregates," *Biofabrication,* vol. 5, p. 015013, 2013.

[22]   R. Devillard, M. Correa, V. Kériquel, M. Rémy, J. Kalisky, M. Ali, *et al.,* "Cell patterning by laser-assisted bioprinting," *Methods in cell biology,* vol. 119, pp. 159-174, 2013.

[23]   D. Gonzalez Rodriguez, K. Guevorkian, S. Douezan, and F. Brochard Wyart, "Soft matter models of developing tissues and tumors," *Science,* vol. 338, pp. 910-917, 2012.

[24]   V. Mironov, R. P. Visconti, V. Kasyanov, G. Forgacs, C. J. Drake, and R. R. Markwald, "Organ printing: tissue spheroids as building blocks," *Biomaterials,* vol. 30, pp. 2164-2174, 2009.

[25] C. Norotte, F. S. Marga, L. E. Niklason, and G. Forgacs, "Scaffold-free vascular tissue engineering using bioprinting," *Biomaterials,* vol. 30, pp. 5910-5917, 2009.

Problems

1. What are autologous cells? Explain their advantages and disadvantages.
2. What are mesenchymal stem cells (MSCs)? List their advantages and disadvantages.
3. What are induced pluripotent stem cells (iPSCs)? Discuss their advantages and disadvantages.
4. Why do cells have to be specifically handled and prepared before bioprinting?
5. What is a cell spheroid? What are its characteristics?

Chapter 6

Three-Dimensional Cell Culture

Over the past decades, three-dimensional (3D) cell culture has been adopted by tissue engineers, cancer researchers, stem cell scientists and cell biologists, mostly in university settings. New materials and methods have continuously been developed. The pioneers of 3D cell culture technology have gained the benefits of better data with extensive knowledge of tissue and cancer behaviour. 3D cell culture methods used to be very expensive, laborious, messy and difficult to adapt to existing procedures. Nowadays, researchers and scientists can select viable 3D cell culture tools to fit their specific requirements and needs. In this chapter, readers will find useful information on the principles and major techniques/tools used for 3D cell culture.

6.1 The Importance of 3D Cell Culture

The current understanding on biological processes is mainly based on studies of homogenous cell populations cultured on flat-surfaces using cell culture flasks, multiwall plates and Petri dishes, usually known as the two-dimensional (2D) static cell culture. This 2D culture typically involves growing single or mixed types of isolated functional cells obtained from human tissues as well as preclinical species on flat plastic or glass substrates. Culture media need to be changed frequently so as to provide fresh nutrients as well as remove metabolic waste. Nevertheless, tissues and organs in the human body are 3D structures and are continuously perfused by the circulation network of the blood. It has been recognised that a significant difference exists in cell functions and behaviours between a flat layer of cells in a medium bath and a 3D

179

complex tissue fed by blood circulation in the body. In fact, tissue-specific architecture, biochemical and mechanical cues as well as cell-cell interaction are lost in 2D culture environment where all the conditions are simplified [1]. Even though certain cellular behaviours can be revealed in a systematic and physiological manner, species' specific behaviours are extremely difficult to be captured by the time-consuming animal models. For instance, idiosyncratic human immunogenicity, toxicity and response from antibodies are usually unpredictable based on animal data [2, 3].

3D cell cultures mimic the *in vivo* situation in which cells spontaneously aggregate to generate tissue spheroids or are embedded into a scaffold to mimic biological environments in living tissues. They can possess key desired features and functions with considerably reduced complexity [4-6]. In 3D culture, proliferation and differentiation in both time and space are regulated by the physiological cell-matrix and cell-cell interactions. Therefore, it is feasible to maintain tissue function and homoeostasis, which has already been proved by several cell biology and proteomic studies [7, 8]. 3D culture models can be applied to studying physiology and pathophysiology of human tissues *in vitro*.

3D culture bridges the gap between 2D cell culture and animal models [9]. Together with human stem cell technology, a wide variety of human tissue models can be created in laboratory environment to study human tissue physiology and pathology, and test drug toxicity, efficacy, safety of consumer products and ingredients.

6.1.1 *3D vs 2D cell culture*

2D culturing, either single or multiple types of cells, is well established and convenient to set up and high viability of cells can be achieved. Cells are seeded onto a culture plastic or glass surface with flasks, multiwall plates or Petri dishes. Culture medium containing necessary nutrients and/or tested chemicals and simulating factors is added to bathe the cells. Cells consume the nutrients whilst producing metabolic waste. As a result, the culture medium has to be changed regularly. When the proliferating cells spread and cover most area of the surface, they must be harvested and subsequently reseeded with reduced cell density. This process is named passaging. Various cell culture equipment (e.g. CO_2 incubator) and consumables (e.g. flasks, multiwall plates and Petri dishes) are available in the market. The simplicity of 2D culture has facilitated researchers and scientists to understand individual cellular activities with substantial reduction of complexity of the native microenvironment in the human body.

However, due to the use of 'over-simplification' concept, the limitation of conventional 2D culture has also been recognised. The availability of standardised consumables, apparatus and instruments makes it difficult to move away from the already well-established methodology. However, cues from 3D environment are required for most cells to form physiological tissue structures *in vitro*. 2D cell cultures are unable to reflect the physiological complexity of real tissue and errors might exist in predicting tissue-specific responses in cell-based assays. Cells cultured in 2D environments mostly undergo proliferation and de-differentiation and as a result, lose their functions [10]. In contrast, cells that are cultured in 3D matrix exhibit significant difference in terms of proliferation, differentiation, functions and morphology [11, 12]. For instance, 2D cultured fibroblasts have a flat shape different from the bipolar shape found in real tissues [13, 14]. On the other hand, fibroblasts cultured in 3D collagen appear to be a more typical *in vivo* phenotype [15]. Primary hepatocytes plated in 2D culturing systems lose viability and liver-specific functions in several days post isolation, however, hepatocytes in 3D collagen matrix or spheroids can maintain high viability and functions such as Phase I and II metabolic activities longer [16, 17]. Moreover, the importance of polarity of epithelial cells can be re-established as spherical *in vivo*-like structures surrounding a lumen in 3D culture models instead of on flat 2D culture substrates [18, 19].

The cell-ECM interactions are bi-directional and a 3D matrix containing variation of density of adhesion-ligand matrix stiffness is capable of guiding cell biological process, and affecting cell functions and morphology [20, 21]. For example, it is feasible to revert the malignant phenotype of epithelial cells of human breast cultured in 3D Matrigel based scaffolds to normal morphology by inhibiting epidermal growth factor receptor (EGFR) and β1-integrin, but not in 2D cell cultures [22].

Mechanical control of phenotype is vital for various adult stem cells at different stages of differentiation. For example, human mesenchymal stem cells can be used to differentiate into multiple tissue lineages by altering substrate compliance: stiffer substrates (8–17 kPa) promote muscle formation [23] and softer substrates (0.1–1 kPa) induce neurogenic differentiation. In addition to offering templates for cells to adhere and grow, 3D cell cultures also provide the interconnectivity within the 3D constructs, enabling nutrients and metabolites to be transported [24]. The high specific surface areas of a 3D porous structure also facilitate a long-term cell culture *in vitro* [25]. Furthermore, well-defined geometry provided by 3D cell cultures enables it to directly relate structure to function. Therefore, data obtained from 3D cultures allows researchers to gain a better understanding of spatial resolution in the body [26-29].

The key features of 2D and 3D cell cultures as research tools for study of living cells *in vitro* are compared in Table 6.1.

Table 6.1. Comparison of key features in 2D and 3D cell cultures.

	Features	2D culture	3D culture
Biological functions	Cell shape (e.g. epithelial cells)	Elongated, loss of epithelial cell polarity	More natural, epithelial cell polarity established
	Morphology	2D substrate induced	3D matrix induced
	Motility	2D substrate dependent	3D matrix dependent
	Cell behaviour and motility	Different to those *in vivo*	More like those *in vivo*
	Cell-cell interaction	Limited	Yes
	Cell-matrix interaction	Limited	Yes
	Stem cell differentiation	2D substrate induced, different to those *in vivo*	3D matrix induced, closer to those *in vivo*
External cell culture manipulation	Culture set-up	Easy	Complex
	Medium change	Yes	Yes
	Continuous culture	Short	Medium
	Microenvironment manipulation	No	Yes

	Mechanical cues (e.g. stiffness and shear stress)	No	Full
Operation and cost	Contamination risk	High	Low
	Culture cost	Cheap	Expensive

6.2 3D Cell Culture Models

Seeding living cells into a 3D scaffold and culturing provides an *in vitro* tissue/organ model that is able to capture important cellular and tissue functions through choosing appropriate scaffold and culturing environment. Model tissues can be generated with varying complexity depending on specific requirements.

6.2.1 *Scaffold-based 3D cell culture*

Continuous efforts have been made on the development of scaffold-based 3D cell cultures for studying specific cell/tissue function *in vitro*. Hydrogels, both natural and synthetic, are particularly useful due to their high water content being beneficial for cell proliferation. In addition, hydrogels are similar to the native ECM of soft tissues from structural and mechanical aspects [30]. For example, alginate has been widely used as a hydrogel ECM for cell transplantation, cell immobilisation and encapsulation, and drug delivery [31-34]. There are many varieties of hydrogels available in the market, such as polyethylene glycol (PEG)-based hyaluronan hydrogels for cancer cells, stem cell and primary hepatocytes study [35], and self-assembling nanofibre hydrogel [36]. Macroporous hydrogel that has tissue-like elasticity has recently been used in 3D cancer cell cultures [37, 38]. In addition, human skin models can be developed by culturing epithelial cells on polymeric fibre mesh or porous membrane. The epithelial cells are normally cultured to confluence at the liquid-air interface on which cell differentiation takes place, generating polarised epithelial sheets [39, 40]. Several methods are available for processing 3D scaffolds, as presented in chapter 2, such as additive manufacturing techniques, freeze drying, micro-fabrication and electrospinning, similar to those used for TE applications.

6.2.2 Scaffold-free multicellular spheroids

Cellular spheroids are formed through spontaneous aggregation of cells from different types. An exogenous matrix or scaffold is not necessarily required to support cells. The widely adopted model is the multicellular tumour spheroids (MCTS), comprised of proliferating cells located on the outside of the spheroid and dead or quiescent cells in the core of the spheroid owing to poor vasculature and the resulting lack of oxygen and nutrient supply [41]. MCTS models mimic avascular micro-tissues with inherent nutrient, oxygen and metabolite gradients. Scaffold-free aggregation of cells can be formed using other types of cells through (i) rotating-well vessels [42]; (ii) cultivation in dynamic cell suspension contained in spinner flasks [43]; (iii) maintenance in cell culture patterned surfaces or inserts; (iv) hanging-drop by gravity-enforced assembly [44]; or (v) liquid overlay method [45]. Biomaterials or ECM are not added onto scaffold-free platforms for spheroid growth. Cells that grow on scaffold-free platforms generate and organise their own ECM and thus, spheroids closely resemble *in vivo* tissues. Co-cultures with other types of cells, namely, stromal, epithelial and endothelial cells, extend the predictive cytotoxicity capabilities of cell culture system. There is no support structure or porosity on scaffold-free platforms. The overall spheroid size is normally limited beyond 500-600μm in diameter.

Spheroid models are primarily applied to cancer pharmaceutical testing and are considered and accepted as an effective approach for toxicity studies due to the spheroid structure and cell behaviour being closer to *in vivo* state in comparison to 2D cell-based models [46].

Despite the unique advantages presented above, some shortcomings arising from spheroid cultures, such as lack of control over the size, need to be addressed. It is possible that cells die in the centre of spheroids during the culture process as a result of insufficient nutrient and oxygen supply which heavily relies on the diffusion from outside to highly dense inner core regions. Dissimilarity to natural tissues and the lack of control of spheroid structure and architecture are the major barriers to the applications of 3D tissue models [47].

6.3 Gels for 3D Cell Culture

Generally, gels have a soft tissue-like stiffness and are aimed at mimicking ECM. Gels produced from ECM mixtures of natural origin e.g. alginate and collagen, have been served as substrates for cell culture. However, residual growth factors, substances (e.g. animal viruses) and undefined constituents are normally contained in gels obtained from animal-sourced natural ECM extracts. Gels are also likely to change in structure over a period of time since they are organised by cells in culture.

Some synthetic gels, such as poly-ethylene glycol (PEG)-based hydrogels, are specifically developed to impart controllability in gels. These hydrogels are modified to obtain desired characteristics. Bulk material modification can be accomplished by hybridisation of natural and synthetic materials, hybridisation of biomaterials with functional nanomaterials and incorporation of different protein types or molecules within the matrix.

A major disadvantage of 3D gels is the difficulty of use due to gelling mechanism. For example, Matrigel^{TM} must always be kept on ice in order to keep its viscosity low enough so that they can be manipulated and mixed with cells. PH-based gelling mechanisms are also very common, whereas, the mechanisms expose sensitive cells to adverse conditions. Other limitations are (i) inherent variability; (ii) impracticality to pour gels in-house for screening; (iii) assessing cells growing in 96-well micro-plates presents the difficulty of maintaining well-to-well consistency; and (iv) difficulties in observing cells in gels without a confocal microscope [40].

6.4 Bioreactors for 3D Cell Culture

6.4.1 *What is a bioreactor*

A TE bioreactor can be defined as a 'device that uses mechanical means to influence biological processes' [48]. Bioreactors can be used to assist the *in vitro* development of new tissue by offering physical and biochemical regulatory signals to cells, and stimulating and encouraging them to differentiate and/or to produce ECM prior to *in vivo* implantation. Bioreactors are devices where biochemical or biological processes develop under a tightly controlled and closely monitored environment. It should be noted that the best bioreactor is the human body. Any *in vitro* bioreactor is only able to replicate a small part of the full functional spectrum of an *in vivo* bioreactor.

6.4.2 *Spinner flask bioreactor*

A spinner flask bioreactor system primarily consists of a media reservoir equipped with side arms, which have porous covers for gas exchange and can be opened to remove media and scaffolds (see Fig. 6.1). There is a stir bar for stirring the media contained in the flask. Needles or thread are used to suspend scaffolds from the top of the flask [49-51]. Typically, scaffolds are fixed and the fluid constantly flows across the scaffold surface, leading to eddies in the superficial pores of the scaffold. These eddies are comprised of fluid particles, which enhances fluid transport. Spinner flasks are usually used in cell cultures for bone tissue engineering since they can increase expression of osteocalcin, alkaline phosphatase and calcium deposition in comparison to static culture [52]. This effect results from the convective nutrient transport to the scaffold surface in spinner flask culture rather than purely diffusional transport which is a common transport manner in static culture. However, mass transfer in flasks is still not fully able to deliver distribution of homogeneous cells throughout scaffolds and most of the cells reside on the construct periphery [53].

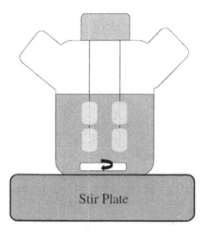

Fig. 6.1. Schematic of spinner flask bioreactor [54] (Copyright 2010, with permission of Elsevier).

6.4.3 *Rotating wall bioreactor*

A rotating wall bioreactor typically has two concentric cylinders, i.e. a stationary inner cylinder for gas exchange and a rotating outer cylinder (see Fig. 6.2). The culture media is filled in the space between these two cylinders and scaffolds are placed anywhere in this space. The scaffolds can move freely, which results in a microgravity environment generating gravitational forces. The forces then balance with the flow of the fluid resulting from the centrifugal forces when the outer cylinder rotates [50, 52]. Moreover, rotating wall bioreactors provide a wider spread distribution of homogeneous cells in comparison to static culture.

Fig. 6.2. Schematic of rotating wall bioreactor [54] (Copyright 2010, with permission of Elsevier).

6.4.4 *Fixed wall bioreactor*

A typical fixed wall bioreactor system contains a stationary culture chamber designed for culturing rigid tissues. The tissues are regenerated through applying specific mechanical stresses. The perfusion solution flows through the tissues and shear stresses are applied onto the cultivated cells by changing the pressure in the culture chamber [55].

6.4.5 *Compression bioreactor*

Compression bioreactors are normally used to culture cartilage tissue and they are designed to allow for both dynamic and static loading. The reason for adopting this kind of design is because static loading has a negative influence on cartilage formation, whereas, dynamic loading – a typical physiological loading – provides better stimulation than many other stimuli [56]. Generally, compression bioreactors comprise a motor, a controlling mechanism and a system providing linear motion. An example of a system is depicted in Fig. 6.3, in which displacements of different magnitudes and frequencies are provided by a cam-follower system. The system can be controlled by a signal generator and the imposed displacement and load response can be measured by using linear variable differential transformers and load cells respectively [57, 58]. Flat platens can distribute loads evenly, which allows the loads to be transferred to the constructs via them. In contrast to static culture, mass transfer is considerably improved in compression bioreactor culture systems since compression leads to the fluid flowing through the scaffolds [57].

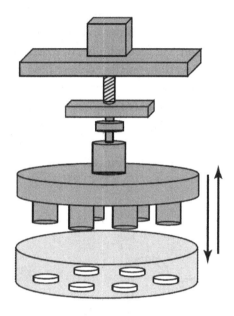

Fig. 6.3. Schematic of compression bioreactor [59] (Copyright 2003, with permission of Elsevier).

6.4.6 *Strain bioreactor*

The designs of tensile strain bioreactors are similar to compression bioreactors and the major difference lies in the way that force is transferred to the construct [60]. This type of bioreactor is used in the culture of various types of tissue including tendon, bone, ligament and cardiovascular tissue. In a strain bioreactor, clamping scaffolds is necessary in order to apply a tensile force. Mesenchymal stem cells are differentiated by applying the tensile strain along the chondrogenic lineage. For multi-station bioreactors, scaffolds can be clamped in uniaxial tension [61].

6.4.7 *Hydrostatic pressure bioreactor*

In a hydrostatic pressure bioreactor, scaffolds are first cultured statically. It is then moved and placed in a hydrostatic chamber. This class of bioreactors is comprised of a culture chamber and a piston for pressurising the chamber [56]. In order to maintain a sterile environment, the piston applies pressure via a membrane, in which case the piston does not have direct contact with the media. Water-filled pressure chamber is a typical variation on this design, which pressurises the chamber via an impermeable film [62]. Figure 6.4 depicts a hydrostatic pressure and perfusion culture system. When filling water into the pressure chamber, the chamber is quickly compressed with an actuator. Hydrostatic pressure is regulated by a backpressure valve connected with the actuator [63].

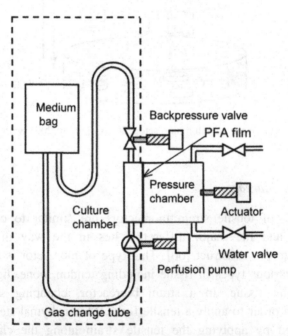

Fig. 6.4. Schematic of hydrostatic pressure and perfusion bioreactor [62] (Copyright 2005, with permission of Elsevier).

6.4.8 *Flow perfusion bioreactor*

Bioreactors that use a pump system to directly perfuse media through a scaffold are known as flow perfusion bioreactors. Most perfusion bioreactor systems consist of a fundamental design, which includes a media reservoir, a tubing circuit, a pump and a perfusion cartridge [64]. Figure 6.5 shows the configuration of a perfusion bioreactor. The scaffold is housed in a sealed perfusion cartridge and thus, media cannot flow around it. The media directly perfuse through the pores of the scaffold. However, such a direct perfusion configuration makes these systems difficult to develop since the perfusion cartridge must be customised for tightly fitting a scaffold and there must be highly interconnected pores on the scaffold. In comparison to other types of bioreactor, flow perfusion bioreactors are the best equipment for fluid transport. If the same scaffold type and flow rate are used, perfusion bioreactors can achieve the highest cell density than other culture methods [53].

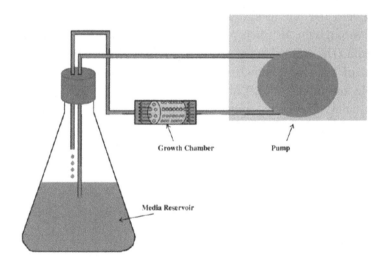

Fig. 6.5. Schematic of perfusion bioreactor [54] (Copyright 2010, with permission of Elsevier).

6.4.9 *Bioreactor with pulsatile pressure*

A pressure bioreactor system primarily includes a pressure piston, semi-compliant tubes, filters, one-way valve, silicon gasket sealants, pressure monitors and a biochamber (see Fig. 6.6). The biochamber is made of polyetherimide, capable of accommodating 6×4 cm sterilisable scaffolds. The biochamber has three access holes inlet pressure, pressure regulation valve and media addition, as shown in Fig. 6.7 (left). For facilitating the media drainage, an outlet is specifically positioned in the lower basement part and is controlled by a valve. The lower and middle parts hold the cell-seeded scaffolds. In order to avoid cyclic deformation to the scaffold, it is placed in between the lower and middle parts, as depicted in Fig. 6.7 (right). Humidified filtered air is taken through the incubator to the pressure pump, which generates cyclical inflow pressure into the bioreactor. A titratable valve is designed to control the pressure regulation, which allows the user to adjust the retained pressure according to the physiologic requirements of the tissue.

When the pressure pump cycles, the pressurised air is mobilised and pushed into the tubing system. The air is pushed into the bioreactor system through the air filter, providing an equally distributed pressure on the media. The exit valve located on top of the biochamber regulates the amount of pressure applied to the bioreactor system. As a result, the pulsatile and unidirectional air pressure is controlled and directed over the vascular constructs, mimicking the pressure in real mammalian vascular physiology.

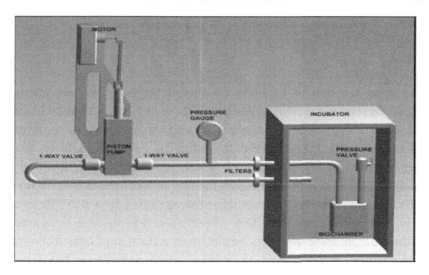

Fig. 6.6. The schematic of a pulsatile pressure bioreactor [65] (copyright 2009, with permission of John Wiley and Sons).

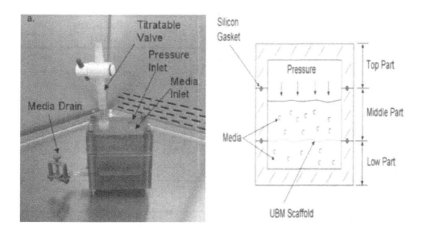

Fig. 6.7. The pulsatile pressure bioreactor [65] (copyright 2009, with permission of John Wiley and Sons).

6.4.10   *Applications of bioreactor on cell culture research*

6.4.10.1   *Bone*

The main function of bones is to support the body, protect internal organs and serve as a mineral reservoir. These highly vascularised tissues consist of spongy bone (inner structure) and compact bone (outer layer). Sikavitsas *et al.* [52] explored the effect of culture vessels on bone regeneration. In the rotating wall vessel, fewer cells were found at a minimal differentiation level due to the lack of mechanical stimuli. Bone differentiation can benefit from shear forces caused by a flow. Bancroft *et al.* [66] investigated the effects of perfusion flow through a construct. A dramatic increase in mineralised matrix production of constructs over static cultures was observed. In addition, ECM and cellular distribution were improved as a result of advantageous fluid shear forces and high nutrient availability. Table 6.2 summarises the principal studies on bone regeneration.

Table 6.2. Bioreactor for bone culture.

Bioreactor	Cell source	Scaffold	Flow	Parameters studied	Ref
Fixed wall, Rotating wall and spinner flask	Rat marrow stromal cells	PLGA	None	Effect of cultivation vessel	[52]
Fixed wall	Rat marrow stromal cells	Titanium mesh	Perfusion	Fluid flow	[66]
Rotating wall vessel	Rat calvarial osteoblastic cells	PLGA	Induced by rotation	Effect of dynamic flow	[67]
Fixed wall	Human fetal osteoblastic cells	Collagen gels	None	Pulsatile uniaxial strain	[68]

6.4.10.2   *Cartilage*

Cartilage is a non-vascularised tissue containing chondrocytes and an ECM composed of glycosaminoglycans (GAG) and collagen. Cartilage is subjected to a wide variety of mechanical stresses and strains. The selected studies are outlined in Table 6.3, showing how cartilage growth can be affected and improved. It is revealed that cellular proliferation is encouraged in medium perfusion by providing efficient transport of nutrients, catabolites, metabolites and gases. The quality of ECM can be improved due to the better uniformity of the pH and mechanical stresses induced by the flow [69]. However, an acidic and anaerobic medium, e.g. intracorporeal conditions, inhibits the ECM formation, which is detrimental for cartilage regeneration [70].

Table 6.3. Bioreactor for cartilage culture.

Bioreactor	Cell source	Scaffold	Flow	Parameters studied	Ref
Fixed wall	Bovine calf glenohumeral joint surfaces	PGA coated by PLLA	0.1-2.0 mL/min	Flow and perfusion	[69]
Fixed wall, cylindrical	Isolated bovine chondrocytes	PLLA	Perfusion	Chondrocyte density	[71]
Fixed wall	Bovine calf articular chondrocytes	Nonwoven PGA mesh	Perfusion	Mechanical environment and perfusion	[72]
Fixed wall	Isolated calf chondrocytes	Agarose hydrogels	Perfusion	Mechanical properties	[73]

6.4.10.3 *Ligament*

Ligaments are fibrous tissues that keep an organ in place or connect the ends of bones. The helical organisation of collagen fibre bundles stabilises the function of the ligament. Mechanical properties of ligaments vary significantly with their location in the body [74]. The research activities in relation to *in vitro* ligament regeneration under both dynamic and static strain are shown in Table 6.4.

Table 6.4. Bioreactor for ligament culture.

Bioreactor	Cell source	Scaffold	Flow	Parameters studied	Ref
Fixed wall	Fibroblasts	Collagen gels	None	Dynamic strain	[75]
Fixed wall	Embryonic chick skin fibroblasts	Collagen gels	None	Relaxed *vs* stretched culture	[76]
Fixed wall	Human bone marrow stromal cells	Silk fibre matrices	Perfusion	Combined dynamic mechanical strains	[77]
Fixed wall	Human lung fibroblasts	Polyurethane nanofibre sheets	None	Direction of mechanical strain	[78]

6.5 Microchips for 3D Cell Culture

Microchips, usually known as 'organ on a chip (OOC)' or microsystems, are considered to be the next promising technology for 3D cell culture models. 'Organ on a chip' incorporates microfluidics technologies with cells that are cultured within microfabricated devices, utilising microchip techniques. The advances in microfluidic and microfabrication techniques open new avenues for the development of 'organ on a chip'.

The recent advances of integrated OOC microsystems have shown their capability in the reproduction of key architectural, functional, mechanical and biochemical features (e.g. shear forces and mechanical strain) of living tissues and organs, including liver, lung, kidney, bone, breast, gut, brain and eye. The ultimate goal is to develop a human-on-a-chip system that contains individual organ microsystems together [79].

OOC technology is currently not affordable and commercially viable. The obstacles to the development and application of this technology are the scale-up and manufacture of microchips. Today's microchips use soft- or photo-lithography and replica moulding techniques together with poly(dimethylsiloxane) (PDMS), silicone rubber and microfluidics systems. These systems are easier to fabricate, less expensive, optically transparent and highly gas permeable. Replica moulding techniques are able to replicate complex surface relief patterns to fabricate biomimetic structures that realistically mimic organ specific microarchitecture. Microfabricated structures are generated within microfluidic channels that resemble the endothelium. Hepatocytes are separated from the liver sinusoid in the entire liver, with the aim of closely approximating the properties of mass transport of the hepatic microcirculation and providing a favourable environment to maintain primary liver cells in a differentiated state [80]. Figure 6.8 shows how a hepatic microarchitecture is reconstituted on a micro-engineered liver-on-a-chip. The functional component of this microsystem comprises a central chamber for culturing liver-cells and a surrounding nutrient flow channel separated by barrier structures that mimic the permeable endothelial barrier between the liver sinusoid and hepatocytes. However, it is noted that PDMS has poor chemical resistance to certain solvents and can absorb tiny hydrophobic molecules. Moreover, the effects of microfluidics and polymers on cellular behaviour remain unknown.

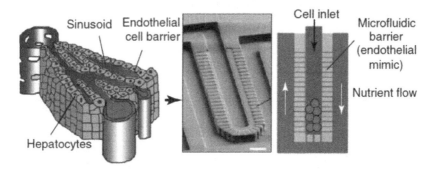

Fig. 6.8. A micro-engineered liver-on-a-chip reconstitutes hepatic microarchitecture [80] (Copyright 2007, with permission of John Wiley and Sons).

6.6 Summary

Table 6.5 summarises the 3D cell culture technologies presented in the previous sections, showing their pros and cons as well as best-suited applications.

Table 6.5. Summary of 3D cell culture technologies.

3D Cell culture technology	Advantage	Disadvantage	Application
Scaffold-based	Large variety of materials available Co-cultures viable Customisable Medium cost	Difficult to remove cells May not be transparent	Cell expansion Drug discovery
Scaffold-free spheroid	Consistent spheroid formation Co-cultures viable No added materials Transparent HTS capable Low cost	Limited flexibility Limited size	Personalised medicine Drug discovery
Gel	Large variety of natural and synthetic materials available Co-cultures viable Customisable Low cost	Gelling mechanism Undefined constituents in natural gels Gel-to-gel variation and structural changes over time May not be transparent	Drug discovery
Bioreactor	High volume cell production Customisable	High cost	Cell expansion Tissue engineering
Microchip	*In vitro* organ specific systems Transparent High gas permeability	Limited HTS options Not commercialised Expertise required High cost	Drug discovery

References

[1] E. Cukierman, R. Pankov, and K. M. Yamada, "Cell interactions with three-dimensional matrices," *Current opinion in cell biology,* vol. 14, pp. 633-640, 2002.

[2] C. Kuperwasser, T. Chavarria, M. Wu, G. Magrane, J. W. Gray, L. Carey, *et al.*, "Reconstruction of functionally normal and malignant human breast tissues in mice," *Proceedings of the*

National Academy of Sciences of the United States of America, vol. 101, pp. 4966-4971, 2004.

[3]   R. Dixit and U. A. Boelsterli, "Healthy animals and animal models of human disease (s) in safety assessment of human pharmaceuticals, including therapeutic antibodies," *Drug discovery today,* vol. 12, pp. 336-342, 2007.

[4]   M. Lutolf and J. Hubbell, "Synthetic biomaterials as instructive extracellular microenvironments for morphogenesis in tissue engineering," *Nature biotechnology,* vol. 23, pp. 47-55, 2005.

[5]   D. Liu, C.-K. Chua, and K.-F. Leong, "A mathematical model for fluid shear-sensitive 3D tissue construct development," *Biomechanics and modeling in mechanobiology,* vol. 12, pp. 19-31, 2013.

[6]   L. Dan, C. K. Chua, and K. F. Leong, "Fibroblast response to interstitial flow: A state-of-the-art review," *Biotechnology and bioengineering,* vol. 107, pp. 1-10, 2010.

[7]   P. Friedl and E. B. Bröcker, "The biology of cell locomotion within three-dimensional extracellular matrix," *Cellular and molecular life sciences CMLS,* vol. 57, pp. 41-64, 2000.

[8]   K. L. Schmeichel and M. J. Bissell, "Modeling tissue-specific signaling and organ function in three dimensions," *Journal of cell science,* vol. 116, pp. 2377-2388, 2003.

[9]   D. Liu, C. K. Chua, and K. F. Leong, "Impact of short-term perfusion on cell retention for 3D bioconstruct development," *Journal of Biomedical Materials Research Part A,* vol. 101, pp. 647-652, 2013.

[10]   K. Bhadriraju and C. S. Chen, "Engineering cellular microenvironments to improve cell-based drug testing," *Drug discovery today,* vol. 7, pp. 612-620, 2002.

[11]   C. Roskelley, P. Desprez, and M. Bissell, "Extracellular matrix-dependent tissue-specific gene expression in mammary epithelial cells requires both physical and biochemical signal transduction," *Proceedings of the National Academy of Sciences,* vol. 91, pp. 12378-12382, 1994.

[12]   S. Mitragotri and J. Lahann, "Physical approaches to biomaterial design," *Nature materials,* vol. 8, pp. 15-23, 2009.

[13]   K. A. Beningo, M. Dembo, and Y.-l. Wang, "Responses of fibroblasts to anchorage of dorsal extracellular matrix receptors," *Proceedings of the National Academy of Sciences,* vol. 101, pp. 18024-18029, 2004.

[14]   S. Rhee and F. Grinnell, "Fibroblast mechanics in 3D collagen matrices," *Advanced drug delivery reviews,* vol. 59, pp. 1299-1305, 2007.

[15]   E. Cukierman, R. Pankov, D. R. Stevens, and K. M. Yamada, "Taking cell-matrix adhesions to the third dimension," *Science,* vol. 294, pp. 1708-1712, 2001.

[16]   M. J. Gómez Lechón, R. Jover, T. Donato, X. Ponsoda, C. Rodriguez, K. G. Stenzel, *et al.*, "Long-term expression of differentiated functions in hepatocytes cultured in three-dimensional collagen matrix," *Journal of cellular physiology,* vol. 177, pp. 553-562, 1998.

[17]   C. E. Semino, J. R. Merok, G. G. Crane, G. Panagiotakos, and S. Zhang, "Functional differentiation of hepatocyte-like spheroid structures from putative liver progenitor cells in three-dimensional peptide scaffolds," *Differentiation,* vol. 71, pp. 262-270, 2003.

[18]   C. M. Nelson and M. J. Bissell, "Of extracellular matrix, scaffolds, and signaling: tissue architecture regulates development, homeostasis, and cancer," *Annual review of cell and developmental biology,* vol. 22, p. 287, 2006.

[19]   K. M. Yamada and E. Cukierman, "Modeling tissue morphogenesis and cancer in 3D," *Cell,* vol. 130, pp. 601-610, 2007.

[20]   D. E. Ingber, "Cellular mechanotransduction: putting all the pieces together again," *The FASEB journal,* vol. 20, pp. 811-827, 2006.

[21]   V. Vogel and M. Sheetz, "Local force and geometry sensing regulate cell functions," *Nature Reviews Molecular Cell Biology,* vol. 7, pp. 265-275, 2006.

[22]   J. B. Kim, "Three-dimensional tissue culture models in cancer biology," in *Seminars in cancer biology,* 2005, pp. 365-377.

[23]   A. J. Engler, S. Sen, H. L. Sweeney, and D. E. Discher, "Matrix elasticity directs stem cell lineage specification," *Cell,* vol. 126, pp. 677-689, 2006.

[24]   S. Basu and S. T. Yang, "Astrocyte growth and glial cell line-derived neurotrophic factor secretion in three-dimensional polyethylene terephthalate fibrous matrices," *Tissue engineering,* vol. 11, pp. 940-952, 2005.

[25]   J. Luo and S. T. Yang, "Effects of Three-Dimensional Culturing in a Fibrous Matrix on Cell Cycle, Apoptosis, and MAb

Production by Hybridoma Cells," *Biotechnology progress,* vol. 20, pp. 306-315, 2004.

[26] A. DeWitt, T. Iida, H. Y. Lam, V. Hill, H. S. Wiley, and D. A. Lauffenburger, "Affinity regulates spatial range of EGF receptor autocrine ligand binding," *Developmental biology,* vol. 250, pp. 305-316, 2002.

[27] S. R. Chary and R. K. Jain, "Direct measurement of interstitial convection and diffusion of albumin in normal and neoplastic tissues by fluorescence photobleaching," *Proceedings of the National Academy of Sciences,* vol. 86, pp. 5385-5389, 1989.

[28] L. G. Griffith and M. A. Swartz, "Capturing complex 3D tissue physiology in vitro," *Nature Reviews Molecular Cell Biology,* vol. 7, pp. 211-224, 2006.

[29] M. J. Saxton, "Modeling 2D and 3D diffusion," in *Methods in Membrane Lipids,* ed: Springer, 2007, pp. 295-321.

[30] M. C. Cushing and K. S. Anseth, "Hydrogel cell cultures," *Science,* vol. 316, pp. 1133-1134, 2007.

[31] A. Garate, A. Murua, G. Orive, R. M. Hernández, and J. L. Pedraz, "Stem cells in alginate bioscaffolds," *Therapeutic delivery,* vol. 3, pp. 761-774, 2012.

[32] T. Garg, O. Singh, S. Arora, and R. Murthy, "Scaffold: a novel carrier for cell and drug delivery," *Critical Reviews™ in Therapeutic Drug Carrier Systems,* vol. 29, 2012.

[33] X. Huang, X. Zhang, X. Wang, C. Wang, and B. Tang, "Microenvironment of alginate-based microcapsules for cell culture and tissue engineering," *Journal of bioscience and bioengineering,* vol. 114, pp. 1-8, 2012.

[34] K. Y. Lee and D. J. Mooney, "Alginate: properties and biomedical applications," *Progress in polymer science,* vol. 37, pp. 106-126, 2012.

[35] G. D. Prestwich, Y. Liu, B. Yu, X. Z. Shu, and A. Scott, "3-D culture in synthetic extracellular matrices: new tissue models for drug toxicology and cancer drug discovery," *Advances in enzyme regulation,* vol. 47, pp. 196-207, 2007.

[36] S. Zhang, F. Gelain, and X. Zhao, "Designer self-assembling peptide nanofiber scaffolds for 3D tissue cell cultures," in *Seminars in cancer biology,* 2005, pp. 413-420.

[37] M. B. Dainiak, I. N. Savina, I. Musolino, A. Kumar, B. Mattiasson, and I. Y. Galaev, "Biomimetic macroporous hydrogel scaffolds in a high-throughput screening format for

cell-based assays," *Biotechnology progress*, vol. 24, pp. 1373-1383, 2008.

[38]   F. M. Plieva, A. Oknianska, E. Degerman, and B. Mattiasson, "Macroporous gel particles as robust macroporous matrices for cell immobilization," *Biotechnology journal*, vol. 3, pp. 410-417, 2008.

[39]   B. A. Justice, N. A. Badr, and R. A. Felder, "3D cell culture opens new dimensions in cell-based assays," *Drug discovery today*, vol. 14, pp. 102-107, 2009.

[40]   F. Pampaloni, E. G. Reynaud, and E. H. Stelzer, "The third dimension bridges the gap between cell culture and live tissue," *Nature reviews Molecular cell biology*, vol. 8, pp. 839-845, 2007.

[41]   J. M. Brown and A. J. Giaccia, "The unique physiology of solid tumors: opportunities (and problems) for cancer therapy," *Cancer research*, vol. 58, pp. 1408-1416, 1998.

[42]   T. Goodwin, T. Prewett, D. A. Wolf, and G. Spaulding, "Reduced shear stress: A major component in the ability of mammalian tissues to form three-dimensional assemblies in simulated microgravity," *Journal of cellular biochemistry*, vol. 51, pp. 301-311, 1993.

[43]   J. Casciari, S. Sotirchos, and R. Sutherland, "Mathematical modelling of microenvironment and growth in EMT6/Ro multicellular tumour spheroids," *Cell proliferation*, vol. 25, pp. 1-22, 1992.

[44]   J. M. Kelm and M. Fussenegger, "Microscale tissue engineering using gravity-enforced cell assembly," *Trends in biotechnology*, vol. 22, pp. 195-202, 2004.

[45]   L. A. Kunz Schughart, P. Heyder, J. Schroeder, and R. Knuechel, "A heterologous 3-D coculture model of breast tumor cells and fibroblasts to study tumor-associated fibroblast differentiation," *Experimental cell research*, vol. 266, pp. 74-86, 2001.

[46]   N. T. Elliott and F. Yuan, "A review of three-dimensional in vitro tissue models for drug discovery and transport studies," *Journal of pharmaceutical sciences*, vol. 100, pp. 59-74, 2011.

[47]   J. Lee, G. D. Lilly, R. C. Doty, P. Podsiadlo, and N. A. Kotov, "In vitro toxicity testing of nanoparticles in 3D cell culture," *Small*, vol. 5, pp. 1213-1221, 2009.

[48]   E. M. Darling and K. A. Athanasiou, "Biomechanical strategies for articular cartilage regeneration," *Annals of biomedical engineering,* vol. 31, pp. 1114-1124, 2003.

[49]   M. Stiehler, C. Bünger, A. Baatrup, M. Lind, M. Kassem, and T. Mygind, "Effect of dynamic 3-D culture on proliferation, distribution, and osteogenic differentiation of human mesenchymal stem cells," *Journal of Biomedical Materials Research Part A,* vol. 89, pp. 96-107, 2009.

[50]   P. Godara, C. D. McFarland, and R. E. Nordon, "Design of bioreactors for mesenchymal stem cell tissue engineering," *Journal of chemical technology and biotechnology,* vol. 83, pp. 408-420, 2008.

[51]   H. J. Kim, U. J. Kim, G. G. Leisk, C. Bayan, I. Georgakoudi, and D. L. Kaplan, "Bone Regeneration on Macroporous Aqueous-Derived Silk 3-D Scaffolds," *Macromolecular bioscience,* vol. 7, pp. 643-655, 2007.

[52]   V. I. Sikavitsas, G. N. Bancroft, and A. G. Mikos, "Formation of three-dimensional cell/polymer constructs for bone tissue engineering in a spinner flask and a rotating wall vessel bioreactor," *Journal of biomedical materials research,* vol. 62, pp. 136-148, 2002.

[53]   A. S. Goldstein, T. M. Juarez, C. D. Helmke, M. C. Gustin, and A. G. Mikos, "Effect of convection on osteoblastic cell growth and function in biodegradable polymer foam scaffolds," *Biomaterials,* vol. 22, pp. 1279-1288, 2001.

[54]   A. B. Yeatts and J. P. Fisher, "Bone tissue engineering bioreactors: dynamic culture and the influence of shear stress," *Bone,* vol. 48, pp. 171-181, 2011.

[55]   R. Sodian, T. Lemke, M. Loebe, S. P. Hoerstrup, E. V. Potapov, H. Hausmann, *et al.,* "New pulsatile bioreactor for fabrication of tissue-engineered patches," *Journal of biomedical materials research,* vol. 58, pp. 401-405, 2001.

[56]   E. M. Darling and K. A. Athanasiou, "Articular cartilage bioreactors and bioprocesses," *Tissue engineering,* vol. 9, pp. 9-26, 2003.

[57]   R. L. Mauck, M. A. Soltz, C. C. Wang, D. D. Wong, P.-H. G. Chao, W. B. Valhmu, *et al.,* "Functional tissue engineering of articular cartilage through dynamic loading of chondrocyte-seeded agarose gels," *Journal of biomechanical engineering,* vol. 122, pp. 252-260, 2000.

[58]   C. Huang, Y. Charles, K. L. Hagar, L. E. Frost, Y. Sun, and H. S. Cheung, "Effects of Cyclic Compressive Loading on Chondrogenesis of Rabbit Bone-Marrow Derived Mesenchymal Stem Cells," *Stem cells,* vol. 22, pp. 313-323, 2004.

[59]   I. Martin, D. Wendt, and M. Heberer, "The role of bioreactors in tissue engineering," *TRENDS in Biotechnology,* vol. 22, pp. 80-86, 2004.

[60]   J. Garvin, J. Qi, M. Maloney, and A. J. Banes, "Novel system for engineering bioartificial tendons and application of mechanical load," *Tissue engineering,* vol. 9, pp. 967-979, 2003.

[61]   L. A. McMahon, A. J. Reid, V. A. Campbell, and P. J. Prendergast, "Regulatory effects of mechanical strain on the chondrogenic differentiation of MSCs in a collagen-GAG scaffold: experimental and computational analysis," *Annals of biomedical engineering,* vol. 36, pp. 185-194, 2008.

[62]   S. Watanabe, S. Inagaki, I. Kinouchi, H. Takai, Y. Masuda, and S. Mizuno, "Hydrostatic pressure/perfusion culture system designed and validated for engineering tissue," *Journal of bioscience and bioengineering,* vol. 100, pp. 105-111, 2005.

[63]   S. Mizuno, T. Tateishi, T. Ushida, and J. Glowacki, "Hydrostatic fluid pressure enhances matrix synthesis and accumulation by bovine chondrocytes in three-dimensional culture," *Journal of cellular physiology,* vol. 193, pp. 319-327, 2002.

[64]   K. Bilodeau and D. Mantovani, "Bioreactors for tissue engineering: focus on mechanical constraints. A comparative review," *Tissue Engineering,* vol. 12, pp. 2367-2383, 2006.

[65]   F. M. Shaikh, T. P. O'Brien, A. Callanan, E. G. Kavanagh, P. E. Burke, P. A. Grace, *et al.,* "New Pulsatile Hydrostatic Pressure Bioreactor for Vascular Tissue-engineered Constructs," *Artificial organs,* vol. 34, pp. 153-158, 2010.

[66]   G. N. Bancroft, V. I. Sikavitsas, J. van den Dolder, T. L. Sheffield, C. G. Ambrose, J. A. Jansen, *et al.,* "Fluid flow increases mineralized matrix deposition in 3D perfusion culture of marrow stromal osteoblasts in a dose-dependent manner," *Proceedings of the National Academy of Sciences,* vol. 99, pp. 12600-12605, 2002.

[67]   X. Yu, E. A. Botchwey, E. M. Levine, S. R. Pollack, and C. T. Laurencin, "Bioreactor-based bone tissue engineering: the influence of dynamic flow on osteoblast phenotypic expression and matrix mineralization," *Proceedings of the National*

Academy of Sciences of the United States of America, vol. 101, pp. 11203-11208, 2004.

[68] A. Ignatius, H. Blessing, A. Liedert, C. Schmidt, C. Neidlinger-Wilke, D. Kaspar, *et al.,* "Tissue engineering of bone: effects of mechanical strain on osteoblastic cells in type I collagen matrices," *Biomaterials,* vol. 26, pp. 311-318, 2005.

[69] D. Pazzano, K. A. Mercier, J. M. Moran, S. S. Fong, D. D. DiBiasio, J. X. Rulfs, *et al.,* "Comparison of chondrogensis in static and perfused bioreactor culture," *Biotechnology progress,* vol. 16, pp. 893-896, 2000.

[70] M. L. Gray, A. M. Pizzanelli, A. J. Grodzinsky, and R. C. Lee, "Mechanical and physicochemical determinants of the chondrocyte biosynthetic response," *Journal of Orthopaedic Research,* vol. 6, pp. 777-792, 1988.

[71] S. Saini and T. M. Wick, "Concentric cylinder bioreactor for production of tissue engineered cartilage: effect of seeding density and hydrodynamic loading on construct development," *Biotechnology progress,* vol. 19, pp. 510-521, 2003.

[72] J. Seidel, M. Pei, M. Gray, R. Langer, L. Freed, and G. Vunjak-Novakovic, "Long-term culture of tissue engineered cartilage in a perfused chamber with mechanical stimulation," *Biorheology,* vol. 41, pp. 445-458, 2004.

[73] C. T. Hung, R. L. Mauck, C. C. B. Wang, E. G. Lima, and G. A. Ateshian, "A paradigm for functional tissue engineering of articular cartilage via applied physiologic deformational loading," *Annals of biomedical engineering,* vol. 32, pp. 35-49, 2004.

[74] G. Vunjak Novakovic, G. Altman, R. Horan, and D. L. Kaplan, "Tissue engineering of ligaments," *Annu. Rev. Biomed. Eng.,* vol. 6, pp. 131-156, 2004.

[75] E. Langelier, D. Rancourt, S. Bouchard, C. Lord, P. P. Stevens, L. Germain, *et al.,* "Cyclic traction machine for long-term culture of fibroblast-populated collagen gels," *Annals of biomedical engineering,* vol. 27, pp. 67-72, 1999.

[76] J. Trächslin, M. Koch, and M. Chiquet, "Rapid and reversible regulation of collagen XII expression by changes in tensile stress," *Experimental cell research,* vol. 247, pp. 320-328, 1999.

[77] G. H. Altman, H. H. Lu, R. L. Horan, T. Calabro, D. Ryder, D. L. Kaplan, *et al.,* "Advanced bioreactor with controlled application of multi-dimensional strain for tissue engineering,"

Journal of biomechanical engineering, vol. 124, pp. 742-749, 2002.

[78]   C. H. Lee, H. J. Shin, I. H. Cho, Y. M. Kang, I. Kim, K.-D. Park, *et al.,* "Nanofiber alignment and direction of mechanical strain affect the ECM production of human ACL fibroblast," *Biomaterials,* vol. 26, pp. 1261-1270, 2005.

[79]   D. Huh, G. A. Hamilton, and D. E. Ingber, "From 3D cell culture to organs-on-chips," *Trends in cell biology,* vol. 21, pp. 745-754, 2011.

[80]   P. J. Lee, P. J. Hung, and L. P. Lee, "An artificial liver sinusoid with a microfluidic endothelial-like barrier for primary hepatocyte culture," *Biotechnology and bioengineering,* vol. 97, pp. 1340-1346, 2007.

Problems

1.   What are the major differences between 2D and 3D cell cultures?
2.   What are the advantages of using 3D culture?
3.   What are the two approaches in 3D cell culture models? Describe their culturing processes.
4.   What are the advantages and disadvantages of using gels for cell culture?
5.   What is the definition of a bioreactor?
6.   List five different bioreactor systems.
7.   If an operation wants to obtain the highest cell density and the highest level of fluid transport, which bioreactor should he/she choose? Why?
8.   Compare spinner flask and rotating wall bioreactors. Identify their advantages and disadvantages.
9.   What are the major obstacles in the development of microchips 'organ on a chip' technology?

Chapter 7

Computational Design and Simulation

Today's computer-aided technologies together with medical imaging have assisted and enhanced the advances in tissue engineering (TE), creating new possibilities in the TE development [1]. Such possibilities include, for instance, using magnetic resonance imaging (MRI) and non-invasive computed tomography (CT) techniques to generate tissue structural images for tissue classification, trauma and tumour identification [2, 3], and three-dimensional anatomical models. These are achieved by using computer-aided design/manufacturing (CAD/CAM) as well as additive manufacturing (AM) technology to manufacture the physical models of hard tissues, scaffolds, and furthermore the customised implant prostheses [4, 5], and applying the physical and anatomical modelling for reconstructive tissue implementation and surgeons [6]. The utilisation of computer-aided technologies has led to the emergence of a new field named computer-aided tissue engineering (CATE) [7, 8]. The advances in biology, biomedicine and information technology for TE application have stimulated the development of CATE [9]. This chapter presents the recent development of CATE with particular interest in computational modelling for bioprinting.

7.1 Tissue/Organ 3D Model Creation

Computer-aided tissue modelling consists of two processes, which are acquisition of imaging data and reconstruction of 3D tissue models. In order to construct a tissue model, anatomic data is first acquired from a proper medical imaging modality. The imaging modality should be able to create 3D visions of anatomy, display the vascular structure, differentiate heterogeneous tissue types, and generate computational tissue models for analysis and simulation.

7.1.1 *Data acquisition, reconstruction and 3D representation*

CT, MRI and optical microscopy are the three imaging modalities that are widely used in tissue modelling. Each of these modalities possesses its own unique advantages but also has inherent limitations, as presented in the following three subsections.

7.1.1.1 *CT and micro-CT (μCT)*

In CT or μCT scans, a sample is exposed to ionising radiation. A density map of the sample is displayed in a number of consecutive 2D images by means of detecting and imaging the absorption of the ionising radiation. Stacking these 2D images generates a 3D view of the scanned area with reasonably high resolution. μCT technology has been used to quantify the relationship between tissues and the designed tissue structures in terms of microstructure and function. For instance, μCT is able to characterise the microstructural and mechanical properties of scaffolds, which facilitates the design and manufacture of customised scaffolds with specific tissue microstructures [10-14].

Differentiation of tissue is accomplished by contrast segmentation. CT is inferior to optical microscopy and MRI (to be introduced in subsections 7.1.1.2 and 7.1.1.3) in terms of differentiating soft tissues with similar densities. On the other hand, CT is far more effective for modelling hard tissues as well as density changes, e.g. the interface between soft tissues and bone. Moreover, μCT technology has been employed in the research of bone density changes for clinical, medical and paleontological applications [14]. The poor performance on the differentiation of soft tissues can be partially attributed to the contrast agents used, most of which are iodine-based and short-lived, and principally of use in imaging the vasculature. Furthermore, there are some metals, metal salts and other metal particulates that considerably increase contrast during CT imaging. Owing to the toxicity of heavy metals that could potentially result in unexpected side effects *in vivo*, the above metallic contrast agents have been used and accepted in clinical applications [15, 16]. A model of an organ blueprint was assembled based on a cadaveric sample, which negated the harmful side effects, indicating that these metallic contrast agents are suitable for clinical applications [13].

7.1.1.2 *MRI*

MRI does not require the sample to be exposed to ionising radiation and as a result, it has been increasingly used in clinical applications. MRI can image soft tissues and bone, showing high superiority in differentiation of soft tissue types as well as recognition and identification of border areas of tissues with similar densities. Similar to CT, a MRI image is obtained by continuously stacking and segmenting a large number of 3D images according to signal intensity. Further segmentation can be realised through selecting relevant voxels with similar value of contiguous signals. This allows models of regions, i.e. part of an individual structure where signal intensities are similar, to be created. For example, a single ligament can be selected rather than displaying all the ligaments in an image.

MRI has been extensively used to assemble anatomic atlases with high resolution. Although the resolution is still lower than that of either CT scans or optical microscopy at present, the resolution of MRI images are expected to increase as this technology matures. For example, Dhenain [17] MRI-scanned mouse embryos and the resolution achieved 20-80μm voxels. The major organs were then isolated by the resulting segmentation.

7.1.1.3 *Optical microscopy*

For developing 3D tissue models, optical microscopic methods require dedicated modelling software to deal with reassembling dissected histology slices. Before an examination, a sample is physically sectioned to 5-80μm in thickness and placed on to slides. The images of the target organ provided by optical microscopy consist of a large number of 2D images that are stacked, precisely arranged and re-aligned in correct positions in the XY plane.

Optical microscopy brings significant superiority over CT and MRI. Individual cells, even diseased tissues, in the body can be visually identified. Stains can be applied to a single slide or all slides. Stains may either be simple (e.g. dyes) or complex (e.g. fluorescing antibodies). Instead of by density or signal intensity in CT and MRI respectively, every tissue type can be differentiated to the definition (resolution) of individual cell in optical microscopy by the features of the cells themselves.

7.1.1.4 *Modality hybrids*

An alternative to obtaining a more precise and accurate 3D tissue model is to utilise multiple modalities for correcting deficiencies in single modality. A typical example is to combine 3D models derived from CT and MRI to display heterogeneous soft tissue. This is because CT is better suited for providing images of bone structures such as skull, and MRI is able to enhance the image resolution. CT and positron emission tomography were combined to identify both metabolic and structural attributes for clinical applications e.g. precisely localising cancer [18].

7.1.2 *Bioinformatics: 2D versus 3D analysis*

A volumetric model comprising of voxels is the basic 3D representation format of successive slice images [19]. The model has surface textures with numerous right angles at each individual voxel scale. There are some reverse engineering methods that have been developed to improve the reconstruction of 3D models which are acquired from a modality in 2D. Additional geometrical information can be added onto the 3D model by conducting Boolean operations. Certain biological units, such as small vessel networks that are too tiny for imaging, can be synthesised to provide more precise information.

Despite the fact that volumetric data (image model) can be directly modified to stereolithography (STL) format for printing on an additive manufacturing machine, there are many benefits when directly using the CAD model that was previously converted from an image model. This is because CAD modelling utilises the 'boundary representation' technique to define a solid model, facilitating the model construction by minimising the file size and ensuring all the bounding surfaces are enclosed.

7.1.3 *3D image representation*

2D segmentation extracts the geometrical CT scan data [20]. Each slice is handled and processed individually, and outer and inner contours of the tissue are detected by using, for example, a conjugate gradient algorithm [21]. The contours are stacked three-dimensionally, which are then used as a reference for generating a solid model typically through skinning operations. Further to 3D segmentation [22] of CT scan data, it is capable of identifying and extracting voxels to form surface images with a high level of geometric accuracy.

3D anatomical image is normally constructed through volumetric representation of segmentation. A volumetric representation involves volume rendering [23] resulting in the surfaces and their voxel-based representation. The representation of the data to be rendered is dealt with in the volume rendering process. Volumetric imaging generates 3D displays with a continuum of image and surface intensity data [24, 25]. Volumetric techniques generate the appearance of 3D surfaces without explicitly defining a geometric surface in the computer. These surfaces are comprised of tiny picture elements i.e. voxels, which are called the basic units of volumetric representation [26].

7.1.4 *Additive manufacturing-based medical modelling*

In the late 1980s, the introduction of AM technologies opens up new possibilities for medical modelling [27]. A CT image is accurately reproduced in several hours as a physical object that can be handled by the surgeon. This allows the surgeon to gain an immediate and intuitive understanding of the complex 3D geometry that can be used to effectively plan and practise an operative procedure.

Steps in the manufacture of a patient model by AM include:

1. Patient scans with CT or MRI imaging.
2. Segmentation to describe and extract the surface as polygons.
3. Generation of a STL file formatted model.
4. Model slicing based on the selected AM process.
5. Model fabrication.

7.2 Scaffold 3D Model Creation

With a wide range of imaging modalities and cell-expression analysis methods (systematic sequencing, proteomics, expression arrays and DNA microarrays etc), bioinformatics establishes key connections within the vast amount of data [28]. The analysis and characterisation of features on the tissue scale can benefit from the understanding of cellular constituents and metabolism. Finer detection and accurate recognition of tissues improve and enhance biological modelling on both anatomic and systematic levels.

7.2.1 *Tissue identification*

A number of tools have been specifically developed for tissue analysis. Nuclei, cell boundaries and ECM elements may be viewed selectively via various optical-microscopy stains.

Tissues are very heterogeneous since cells sometimes express complex transitional phenotypes. Some outlier cells cannot be recognised by pattern-based recognition systems. These outlier cells might include immune cells migrating through the tissue, mesenchymal stem cells, cancerous cells, or cells whose nucleus or other features did not stain properly. When a couple of cells appear to be different from the surrounding cells, the data can be augmented with K-nearest neighbour classification [29]. This pattern recognition method classifies the anonymous cell as identical with the rest of the neighbouring cells.

Once the images are obtained, an automated method is used to identify and separate the tissue types and the individual cells respectively, depending on the intended scale of the reconstruction. For CT and MRI, differentiation of tissues is performed by contrast segmentation. Different types of tissues are separated according to the values of signal intensity. The optical approaches operating in RGB (red, green and blue) colour offers triple discriminatory capability of grey-scale. Further methods may also be applied, such as region growing. Although tissue separation can be accomplished to some extent by conducting contrast segmentation, the identification of tissues requires fundamental knowledge-based methods.

7.2.2 Analysis of cells

There are many automated methods that can be used for cell counting, geometry determination, chromosomal counting and determination of correlation of DNA expression. All these features possess predictive value for tissue viability determination in cell culture. DNA expression provides a description with great detail of cell function, identifying up to thousands of expressed genes at one time.

In DNA microarrays, thousands of DNA sequences are printed on the glass slides. When placing a cell sample onto a slide, it will visibly interact and share expression with the relevant DNA spots. Each spot is then automatically and visually detected and the intensity is measured, followed by cataloguing the type of genes [30]. In comparison to DNA microarrays, immunoassays provide even greater specificity since they can be employed to identify virtually every cellular component or expression element. However, sample sizes have to be large enough for this method. Additionally, the spot analysis should be highly automated to expedite results for DNA microarrays and immunoassays modalities [31]. Due to the large data quantity, the data clustering by correlation should be performed without supervision in most cases.

7.2.3 Anatomic registration

In addition to the recognition of individual cells in a selected tissue region, it is equivalently important to automatically recognise organs. The data for conducting organ recognition are from MRI and CT. The recognition of organs is able to robustly and quickly separate out defects and organs at the anatomic level, notwithstanding individual patient variation.

There are two popular automated organ recognition methods, namely, feature-based and shape-based. The feature-based methods detect features or landmarks, e.g. identification of the major vessels of a kidney. The shape-based methods compare the imaged date with an identified shape with some margins for error. The fundamental difference between feature and shape-based methods lies in the shape-based method that takes into account Euclidean information for features and measures structural correspondence, which shows 2-33% accuracy improvements over feature-based methods [32].

7.3   Computer-Aided   Tissue   Scaffold   Design   and   Manufacturing

The design of tissue scaffolds for TE applications requires careful considerations on the complex hierarchy as well as structural heterogeneity of the scaffold and the host tissue. In addition to important factors such as pore size, porosity, interconnectivity as well as transport property for oxygen and nutrients, the designed and manufactured scaffolds should possess compliant mechanical properties related to the host environment [33, 34]. For example, sufficient mechanical strength is required for load-bearing scaffolds or substitutes used in implantation e.g. bone and cartilage [35, 36].

7.3.1   *Biomimetic modelling*

The load-bearing scaffolds have certain characteristics that enable them to work and function as a substitute e.g. bone substitute that meets the mechanical, biological and anatomical requirements. Such characteristics include the following:

- Mechanical requirement: the scaffold should be able to provide sufficient structural support at the replacement site where the desired tissue will regenerate. As a result, the produced scaffold structure should exhibit the required mechanical strength and stiffness.
- Biological requirement: the scaffold must enable cell attachment and distribution, regenerative tissue growth and transport of nutrients and signals. This requires precise control of the structural porosity, by selecting appropriate biocompatible materials and providing decent interconnectivity inside the scaffold structure.
- Anatomical requirement: the geometric size of a tissue scaffold must be appropriate to allow it to fit in at the replacement site [37].

In general, a CATE-based approach begins with non-invasive image acquisition and image processing of the selected tissue regions [38]. This is followed by:

(1) an anatomical structure reconstruction using reverse engineering and image reconstructive techniques;
(2) computer-aided design of scaffolds to represent a variety of tissue morphological and anatomical features;
(3) characterisation of the mechanical properties and structural heterogeneity of designed unit cell and tissue by using the CT technique in order to choose suitable unit cells for the final tissue scaffold;
(4) finalising the scaffold design with the specified anatomic compatible external geometry and internal architecture.

The subsection below provides a detailed example of the design of a bone scaffold using a CATE-based biomimetic modelling technique.

7.3.1.1   *CATE-based biomimetic modelling of bone scaffolds*

Step 1: image acquisition

In Sun *et al.*'s study [39], the CT images were first obtained from that of a proximal femur bone. In total, a total length of 68 mm of the proximal femur bone was obtained consisting of 34 sliced images. The sliced images were then loaded into the software named MIMICS developed by Materialise®, where they were organised sequentially and oriented for the top, bottom, anterior and posterior positions.

Step 2: segmentation and characterisation process

Once the sliced images were loaded, the next step was to identify the region of interest and create a 3D voxel model to study. The relevant information contained in the femur was captured by setting up an appropriate threshold range. All pixels that were within the defined range were gradually grown to a colour mask and as a result, the segmentation was achieved by using the region-growing techniques embedded in the system.

Step 3: CAD model generation

Upon characterising the bone structure, a complete CAD database containing information of the structure was to be generated. 3D reconstruction was conducted to obtain a CAD-based tissue model and three interfaces with CT/MRI data were used [39]: (i) MedCAD interface (this is an interface with medical-imaging data in MIMICS); (ii) reverse engineering interface; and (iii) STL interface that uses STL file in MIMICS. The overall procedure for this CATE biomimetic modelling and the bone tissue scaffold design is illustrated in Fig. 7.1.

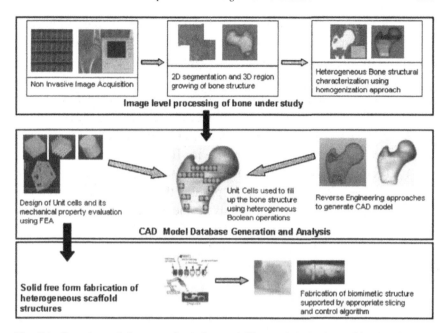

Fig. 7.1. Overview of the procedures for modelling and designing a biomimetic bone scaffold [39] (Copyright 2004, with permission of John Wiley and Sons).

7.3.2 *Design for tissue scaffolds*

7.3.2.1 *Overview*

In order to build TE scaffolds of high quality for the human body, two major challenges lie in the customisation of scaffold pore size and porosity for different cell types, and the development of the scaffold's mechanical, biological and anatomical properties mimicking the native tissue [40, 41]. Additionally, the external geometry of the scaffold should fit into the site of the defect since it will determine the stress distribution at the interface of organ implant and the surrounding tissue [42, 43].

The general approach of CATE for scaffold design can be summarised as follows [7, 9, 44]:

1. Acquiring the image of the defect from medical imaging techniques e.g. CT or MRI to obtain the geometry and the form of the defect to be regenerated or repaired.
2. According to the needs of the defect site, selecting appropriate shapes of unit cells from a library storing basic shapes of unit cells for the internal architecture of the scaffold [33, 45].
3. Stacking up the cellular units to construct a block, of which the size is larger than the defect size for Boolean operation.
4. Performing Boolean operation between the stacked cellular units and the defect image after placing the defect image over the arranged stack of cellular units for the generation of the form and architecture of the scaffold [46].

7.3.2.2 *Scaffold design architectures*

(*I*) *Periodic porous structures:*

For native tissues, their physiological structures are very complex and heterogeneous. As opposed to exactly reproducing their internal microarchitecture, scientists and researchers largely concentrate on the fabrication of simplified models, of which the mechanical properties and porosity are functionally equivalent to those of the tissue that is needed to be repaired. With this purpose in mind, scaffolds produced by AM techniques have been designed by the use of different elemental/modular units that exhibit well-characterised transport and mechanical properties. Indeed, beginning with a few repetitive units, various scaffolds with tuneable properties and architectures can be designed in a short period of time.

A number of different methods for scaffold designs have been specifically developed for different scaffold fabrication techniques. For laser-based AM systems (e.g. SLS and SLA) and printing techniques (e.g. colour jet printing), design methods mimic tissue architecture by means of 3D unit cell tessellation, creating unit cell libraries (an example is given in Fig. 7.2) at different scales. The unit cells can then be arranged and assembled to generate the scaffold with complex patient-specific structures [47]. However, it should be noted that these methods are not applicable to extrusion AM techniques such as fused deposition modelling (FDM) since they build the interior structure primarily consisting of continuous and regular patterns of rod-shape elements. The design methods for regular porous scaffolds are described in the following text. Architectures obtained by using periodic unit cell replication are referred to as 'cellular structures'. Architectures comprised of crossing laths, rods or other strips of material are entitled 'lattice domains'.

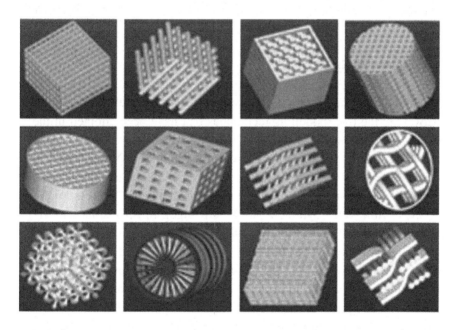

Fig. 7.2. A library of unit cells based on primitives [39] (Copyright 2004, with permission of John Wiley and Sons).

Image-based design:

The image-based design method was originally developed by Hollister *et al.* [48] for the design of site- and patient-specific scaffolds. Image-based design combines imaging and image processing with AM techniques to simplify the procedures in scaffold designs. This method generates scaffold microarchitectures by intersecting two 3D binary images. One of the images represents the shape of the defect, which is acquired from medical imaging data, and another one is obtained by stacking binary unit cells.

By adopting this design method, both bio-inspired and empirically derived geometries can be generated. This method has already been used to design and manufacture scaffolds for craniofacial reconstruction e.g. orbital floors, mandibular condyles and generic mandibular defects for *in vivo* tests [49, 50]. Moreover, combining the image-based method with topological optimisation algorithms enables scaffolds to be designed with conflicting requirements on mass-transport and functional properties [51].

CAD-based methods:

Boundary representation (B-Rep) and constructive solid geometry (CSG) are the two modelling techniques that have been adopted in most of the commercial CAD tools. In CSG-based design systems, scaffold models are designed using standard solid primitives. Nevertheless, the design capability is largely restricted by the limited solid primitives, typically represented by simple and basic geometric objects such as spheres, cubes and cylinders. In B-Rep algorithms, a solid object is presented through its boundaries, comprising a set of edges, vertices and loops, without explicit relation between them. As a result, a preliminary check is compulsory to ensure that there are no overlaps or gaps among the boundaries [52]. In addition, it is noted that, in comparison to CSG models, B-Rep ones require less computation time but more space for data storage. As a result, the file size significantly increases as the object contains fine internal architectures or becomes large, leading to difficulties in visualising and manipulating the objects.

The popular commercial CAD software based on CSG or B-Rep modelling principles are CATIA (Dassault Systemes), NX (Siemens PLM Software), SolidWorks (Dassault Systemes), Pro/Engineer (PTC) and MIMICS (Materialise®). These pieces of software dominate the CAD market and they are the most frequently used ones in the design of 3D structures [53, 54]. Many CAD-based units that are derived from different primitive geometries have been characterised and combined into scaffold microarchitectures. Moreover, a number of polyhedral shapes including Archimedean and Platonic solids have been studied and chosen to be basic building blocks for constructing wireframe approximations [55, 56].

Comparing image-based with CAD-based methods, the former one facilitates rapid design and generation of scaffold architectures as a result of their good compatibility with MRI and CT imaging. However, the image-based method is limited by the dataset dimensions [7]. CAD-based methods are advantageous in generation of multiple datasets with varying resolutions for the same design of the scaffold, which facilitates the use of different length scales for designing the interior porous architecture (micro) and the external anatomical shape (macro).

Implicit surfaces and space-filling curves:

Implicit surface modelling (ISM) is a flexible method for generation of cellular structures, which provides a compact representation of complex surfaces [57, 58]. Scaffold architectures can be easily described using a mathematical equation, with the ability to introduce different architectural features such as pore size gradients.

The design methods reported above can barely be coupled to extrusion-based AM techniques [59]. Take the FDM process as an example. Given the inherent manufacturing constraints, researchers have been looking for patterns that (1) do not require the start and stop points (to avoid material agglomeration due to delayed response time) and (2) do not cause filament intersection. The use of a repetitive pattern in the design of scaffolds has also been considered to be a factor to simplify the deposition process. Hence, the simplest scaffold that can be produced by FDM has cube-shaped pores created by orthogonal rasters. More complex patterns (e.g. honeycomb-like patterns) can be generated by simply changing the angle of deposition between adjacent layers [60] (See Fig. 7.3A). 3D scaffolds with controlled porosity and required mechanical properties can be obtained by varying architectural parameters such as spacing, fibre diameter and stacking direction [61, 62].

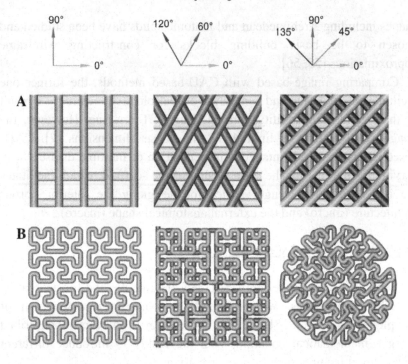

Fig. 7.3. Print patterns with (A) honeycomb pores and (B) Hilbert curves [7] (Copyright 2013, with permission of Elsevier).

Although a considerable number of scaffolds have been fabricated using the above mentioned patterns and extremely successful results have also been obtained, low efficacy of cell seeding and heterogeneous cell distribution are usually considered as the major disadvantage of extrusion-based techniques. A method named space-filling curves has been proposed as an improvement to simple patterns for designing scaffold internal architectures (see Fig. 7.3B). These curves fill up the given space by either passing through every point in the defined space or recursively changing trajectory [63]. Space-filling curves can be mathematically generated by beginning with a simple pattern that grows through the recursive application according to a small set of rules defined by a generative grammar.

Space-filling curves have been proved to be applicable to TE scaffold design by Starly and Sun [64]. Two space-filling curves i.e. Sierpinsky and Hilbert were separately used for the manufacture of circular scaffold by precision extrusion deposition, providing a substrate for cell organisation. Space-filling curves were further applied to the scaffold design for bone TE with subject-specific external geometries. The material gradients within selected scaffold regions and the varying porosity can also be obtained by changing the curve type and modifying its iteration level [65].

(II) Irregular porous structures:

The designs using periodic porous structures are difficult to locally control pore size, shape and distribution due to regularity and periodicity. A minor modification of the unit cell may result in global changes to the entire structure. In addition, CAD tools currently available are not fully capable of reproducing natural structures with complex arrangement [66].

Therefore, biomimetic design has been proposed and extensively used as an effective alternative for modelling irregular porous structure. The complex microstructures of some natural materials such as bone and wood are known to be optimised for bearing functional loads. Thus, the synthetic scaffold designs inspired by high-performance structures have received increasing interest [67]. Biomimetic models accurately mimicking the interior configuration of natural bone can also be developed by starting from histological data and computer tomography [68]. It should be noted that, in most cases, a faithful reproduction is not strictly necessary but the key functional properties of natural architectures are still required to be caught in an efficient and concise manner. Hence, a simpler approach to achieving a biomimetic design is mimicking the functionality of tissues and organs by utilising varying porosity at different regions in accordance with a natural reference model. Observations of tissues such as skin and bone reveal that these structures show well-organised architectures in terms of anatomical, biological and mechanical characteristics. Scaffolds consisting of material gradients are termed functionally graded scaffolds (FGSs) [34].

The design of FGSs requires new developments on computer-aided systems to obtain the desired porosity distribution in an interconnected and continuous way throughout the entire scaffold geometry. Disconnectivity in terms of deposition tool path planning and discontinuity at the interface of two neighbouring regions are usually detected in scaffolds with varying porous architectures, which are detrimental to nutrients, fluid flow and waste transport [69]. Based on the above concerns, a number of strategies have been developed to optimise scaffold design with appropriate continuous functional gradients. Differently spaced tool paths can be connected to improve connectivity and continuity between different functionally graded regions [70]. CAD-based unit cells are also utilised in the design of FGSs. The libraries of unit cells are created where the connection between unit cells are handled by introducing common interfaces. A common connect feature, namely, a torus shape, is added at each cellular facet for the polyhedral units in order to generate a smooth change between different microarchitectures. Furthermore, Voronoi and stochastic models have been applied to mimicking the randomness of natural tissues and then translating it into scaffold design. By implementing stochastic methods, scaffolds with heterogeneous pores distributed biomimetically according to a defined porosity level can be designed and manufactured [71, 72].

7.3.2.3 *Design optimization strategies*

A 'trial-and-error' method has been employed to validate scaffold microarchitectures for decades. Ex post modifications will be made to an existing design based on *in vitro* or *in vivo* results. Additionally, *in silico* experiments facilitate optimisation of design parameters and identification of suitable configurations for replacing desired tissues [73].

A few methods have been proposed to model scaffolds with intricate internal architectures and subject-specific external shapes. For example, in analytic models, global mechanical behaviour can be estimated by analysing empirical relationships between mechanical properties and structural parameters. Finite element analysis (FEA) technique is applied to the investigation of scaffold properties for modifying architectural parameters based on specific requirements. Although CAD designs provide accurate models of scaffold morphology, large discrepancies still exist between the intended design and the final AM fabricated objects due to the inaccuracy of the AM process (dimensional accuracy, surface roughness, micro-pores etc.). The influence of these discrepancies still remained poorly understood [74, 75]. As a consequence, FEA has been chosen in the μCT reconstructions of AM manufactured scaffolds, rather than completely relying on CAD models [76, 77]. Successful results have been obtained in the prediction of mechanical properties of AM manufactured polymeric and composite scaffolds [78, 79]. Diffusivity is a rather important factor in the evaluation of a scaffold design. Diffusivity refers to a function of porosity-related parameters e.g. pore tortuosity and volume fraction. The integration of the FEA technique in the calculation of effective diffusivity offers a rigorous method to determine the diffusion of nutrients and oxygen as a function of scaffold microstructure [80]. FEA has also been combined with empirical investigations to study the impact of matrix degradation on tissue ingrowth, evaluating the effects caused by the degradation process, particularly in terms of mechanical functionality [81, 82].

A permeable pore network facilitates scaffold wetting and improves transport of oxygen and nutrients, whereas, it cannot guarantee adequate and long-term cell viability. It has been recognised that static seeding and culturing, widely used in laboratory environments though, are characterised by inhomogeneous cell distribution and low seeding efficacy. In addition, cell adhering to the scaffold surface consumes a large amount of oxygen and nutrients, resulting in cells in the interior struggling to survive under the condition of oxygen deprivation and hypoxia-induced death [83]. Therefore, medium perfusion has been suggested for increasing the cell survival rate within porous scaffolds, leading to more homogeneous tissue engineered constructs. This stimulates the development of perfusion bioreactors where a fluid continuously flows through the scaffold pores, enhancing mass transfer [73].

In addition to FEA, computational fluid dynamics (CFD) is adopted to investigate the permeability of AM scaffolds for optimisation of scaffold geometries used in skeletal tissue engineering [84]. The use of CFD has enabled the design of optimised scaffold microstructures in terms of mass transfer of gases and nutrients. CFD models have been used to investigate the local shear stress distribution within scaffolds [85-87] and the influence of scaffold micro-architectural parameters [88, 89].

Moving beyond the computation of scaffold properties, the combination of *in silico* modelling with topological optimisation has become an effective method for the *a priori* design of optimised scaffold architectures [74]. In the design of scaffolds, both conflicting requirements (i.e. mass transport and mechanical properties) have to be addressed. Mass transport can be increased by designing large scaffold porosity, whereas, increasing porosity negatively affects mechanical properties. New design strategies are needed to find the balance and the optimal trade-off between the two opposite needs whilst satisfying both biological and mechanical requirements. Topology optimisation aims to utilise material appropriately within a structure subjected to either only single or multiple load distribution [90]. The optimisation methods involve the determination of features, such as the number and shape of pores, their locations and connectivity of the domain. Thus, the design of scaffolds relies neither on a predefined library of unit cells, nor designers' ability. These methods have been used to optimise mechanical properties with a porosity constraint [91] and to maximise scaffold permeability [92]. The most successful application of the topology optimisation is the design of microstructures with optimised permeability for mass transport and cell migration with mechanical properties comparable to those of natural bone tissues [51].

7.4 Case Studies

7.4.1 *Computer-aided System for Tissue Scaffolds (CASTS)*

Computer-aided system for tissue scaffolds (CASTS) developed by Naing *et al.* [93] and Sudarmadji *et al.* [94] is an in-house parametric library of polyhedral units that can be assembled into customised tissue scaffolds. The platform of CASTS is a 3D CAD/CAM system developed based on Pro/ENGINEER. CASTS possesses three modules, namely, the input module, the designer's toolbox and the output module. Thirteen polyhedral configurations are available for selection, depending on the biological and mechanical requirements of the target tissue or organ. Input parameters include the individual polyhedral units and overall scaffold block as well as the scaffold strut diameter. Taking advantage of its repeatability and reproducibility, the scaffold file is then converted into a STL file and fabricated using selective laser sintering, an AM technique. CASTS seeks to fulfil anatomical, biological and mechanical requirements of the target tissue or organ. Customised anatomical scaffold shape is achieved through a Boolean operation between the scaffold block and the tissue defect image. Biological requirements, such as scaffold pore size and porosity, are unique for different type of cells. Matching mechanical properties, such as stiffness and strength, between the scaffold and target organ is very important, particularly in the regeneration of load-bearing organs, i.e., bone. This includes mimicking the compressive stiffness variation across the bone to prevent stress shielding and ensuring that the scaffold can withstand the load normally borne by the bone. The stiffness variation is tailored by adjusting the scaffold porosity based on the porosity–stiffness relationship of the CASTS scaffolds. Two types of functional gradients based on the gradient direction include radial and axial/linear gradient. Radial gradient is useful in the case of regenerating a section of long bones while the gradient in linear direction can be used in short or irregular bones. Stiffness gradient in the radial direction is achieved by using cylindrical unit cells arranged in a concentric manner, in which the porosity decreases from the centre of the structure toward the outside radius, making the scaffold stiffer at the outer radius and more porous at the centre of the scaffold. On the other hand, the linear gradient is accomplished by varying the strut diameter along the gradient direction. The parameters to vary in both gradient types are the strut diameter, the unit cell dimension and the boundaries between two scaffold regions with different stiffness.

7.4.1.1 *Input model*

CASTS uses imaging software, i.e. MIMICS for its input module. This imaging software is used for converting raw patient data from MRI CT scans to surface files of initial graphics exchange specification (IGES) format, which is a neutral data format for file transfer to dissimilar system.

7.4.1.2 *Designer's toolbox*

The designer's toolbox is to provide the CASTS user with a variety of useful tools to customise the scaffold according to patient's specific needs. The tools include the following:

- Parametric library, where the user can select a basic cellular unit/packing configuration, which will be the basis of the scaffold block. Besides that, the user can also decide whether the presence of triangulation or node is required.
- Sizing routines, where the user can specify the dimension of each cellular unit and also the overall dimension of the scaffold block in the x, y and z directions.
- Automated algorithm to link the choices/values inputted by the user and the CAD file of the scaffold. Besides generating the scaffold according to the user's specified size, necessary outputs, such as porosity, pore size and surface-to-volume ratio (scaffold internal architecture data), are also generated.
- Scaffold internal architecture data to allow the user to obtain critical information about the scaffold generated, such as porosity, pore size, and surface-to-volume ratio. These values are calculated based on the dimensions given by the user.

7.4.1.3 *Output model*

The output of the designer's tool module is a scaffold block in the PRT file format. To customise the scaffold to the patient's specific shape, the surface files of IGES format from the input module are then used. After the customisation is done, the part file must be converted into STL file format for scaffold fabrication using SLS.

CASTS system consists of polyhedral unit cells that can be assembled together to form a scaffold block. The polyhedra used in CASTS are convex polyhedra, whose shapes are not too complex in order to be easily modelled in CAD environment. The assembly process was done by repeating the unit cells and mating their faces with the faces of their adjacent unit cells. Some of the polyhedral unit cells with the resulted scaffold blocks are shown in Fig. 7.4.

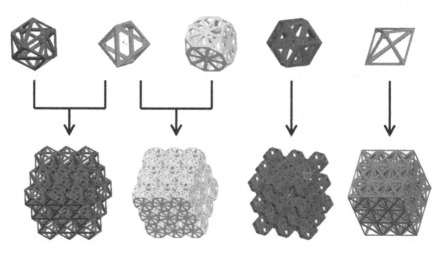

Fig. 7.4. Five of the polyhedral unit cells with the resulted scaffold blocks obtained from combining the unit cells [94].

7.4.2 *Computer-aided Tissue Engineering of the Anterior Cruciate Ligament Repair*

This section presents the application of CATE for the design of a clinically relevant scaffold for anterior cruciate ligament (ACL) repair [95]. Textile architecture, i.e. an assembly of concentric circular braded fibres/layers [96] in this case study, was selected as the architecture of the scaffold and copoly(lactic acid-co-(ε-caprolactone)) (PLCL) was chosen as the scaffold material. The influence of the process parameters on the morphology and tensile response of the scaffold were quantified, as presented in the following two subsections.

7.4.2.1 *Computer-aided morphological characterisation*

The morphology characterisation was based on a numerical description of the scaffold obtained from the fibre trajectory analysis during the braiding process. For each braided layer, the fibre trajectories were interpolated from discrete positions of the fibre cross-section centres. The 3D braid was created assuming a constant fibre diameter. The multilayer braided geometry was generated by repeating this process with differing reference diameters, and by iteratively decreasing the distance between the present layer and the previous one. This virtual scaffold geometry was validated through comparing its cross-sections with those of actual scaffolds, and specifically by computing radial and angular density distributions [97]. The morphology was characterised by the distribution of the pore sizes within the virtual scaffold and the pore interconnectivity. The validity of the computed distribution of the pore sizes was confirmed by using mercury porosimetry [98]. It was revealed that the pore size distribution, in terms of maximum, median and mean pore size, is significantly affected by the diameter of the fibre and the number of layers in the scaffold. The designed multilayer braided architecture fulfilled the requirements for tissue engineering of ligament in terms of pore size and interconnectivity.

7.4.2.2 *Mechanical modelling at the fibre scale*

Anterior cruciate ligament is constantly subjected to stretch and torsion during daily activities and thus, the macroscopic loading is the predominant cause of the mechanical stimuli detected at the cellular scale [99]. The FEA modelling of the dry scaffold was carried out which assumed that biological fluid did not have significant effect on the macroscopic response of the scaffold and the mechanical stimuli.

The nonlinear behaviour of textile materials was reproduced by simulating the relative motions between fibres, independently modelled by a kinematical beam model describing cross-sectional strains within a deformation framework. An elasto-plastic constitutive law was applied to the approximation of the non-linear stress-strain response of PLCL fibres. The contact-friction interactions were simulated in the form of a

Coulomb's law and a regularised penalty law. FEA started from an arbitrary configuration in which fibre trajectories were described by helices and thus, fibres were interpenetrated within the same layer. The interpenetration was reduced gradually through contact conditions starting with the braiding pattern of the construct. A non-interpenetrated configuration of braid was first obtained as the equilibrium configuration of fibre assembly. Having developed this initial configuration, a number of boundary conditions were prescribed, which was aimed at reproducing the braiding tension required to tighten fibres during the braiding process. The layers can freely be rearranged with each other. The torsion and tensile tests were then simulated by prescribing an increasing rotation and displacement to one edge of each layer, respectively. The developed FEA model was capable of predicting the nonlinear response of the braided construct in terms of yield load and stiffness with a slight overestimation of the stiffness as compared with the actual scaffold stiffness.

7.5 Computational Modelling for Bioprinting

7.5.1 *Importance of Computational Modelling for Bioprinting*

In tissue engineering, one of the fundamental theoretical and practical interests is to understand the growth of living tissues [100]. Evolution of both animal and plant tissues can be predicted by using analytical models of tissue growth, which can improve the treatment of pathological conditions as well as offering new prospects in tissue engineering. However, biological and biochemical mechanisms of tissue growth have not been thoroughly understood. Biochemistry, obviously, is the driving force for tissue growth and the understanding of biochemistry of growth becomes highly desirable. Although biochemistry can be used to explain why a tissue grows, it is still necessary to understand how the tissue grows. In order to do that, a macroscopic description in terms of the parameters that can be macroscopically measured is required. There have been plenty of macroscopic models of soft tissue growth [101-104]. Nevertheless, the mathematical apparatus of the existing methods is fairly complicated, which involves variables that are rather difficult to interpret in simple terms and to measure in experiments. This requires continuous efforts on developing comprehensive models.

Another important consideration that requires computational simulations is tissue fusion, which is a fundamental phenomenon during embryonic development and morphogenesis [105]. In bioprinting, there are three main sequential steps, which are the design of organ 'blueprint' (pre-processing), the actual printing process (processing) and accelerated tissue maturation with specific bioreactors/incubators (post-processing) [106]. During the incubating process, the construct comprised of the cellular aggregates immersed in a hydrogel will morph into a 3D tissue according to the natural rule of histogenesis and organogenesis [107, 108]. Tissue self-assembly, fusion, proliferation, differentiation, and maturation take place. Tissue fusion driven biofabrication is an essential biophysical and biochemical process in the bioprinting technology. So far, however, the biophysical and biochemical mechanism of tissue fusion has not been well understood due to the complexity of cell motility. Modern computer simulations and mathematical modelling open up the opportunity to explain the nature of tissue fusion process and thereby plausible outcomes can be predicted. This facilitates the design and optimisation of bioprinting processes. Computational modelling incorporates an on-board programming language, together with physics simulation, animation and visualisation capabilities. A computational model is considered to be a supplement for experimental efforts to rationalise the workflow of matrix design.

7.5.2 The development of computational models

Starting from the early 20th century, modelling of living tissues has evolved along two conceptual lines. In the first concept, a tissue is considered as a set of discrete and interacting cells [109]. By contrast, the other concept treats a tissue as a set of continuum cells and cell densities are monitored instead of individual cells [110]. The following briefly introduces a few of these theoretical models.

The continuum approach, proposed by Murray *et al.* [111], employs the methods of continuum mechanics and considers modelling tissues comprised of realistic numbers of cells. Cell density is used to describe the distribution of cells with various types throughout the entire tissue. In the meantime, the morphogenetic rearrangements are considered as fluxes. The method has been used to investigate developmental morphogenesis, scar formation, dermal wound healing, contraction and vasculogenesis [110].

Growing vascularised large organ replacements in a laboratory environment is a challenging task for tissue engineers. Thus, understanding the detailed mechanisms of vasculogenesis is crucial for creating perfusable tissue constructs. The model developed by Murray *et al.* [110] is able to predict the emergence of filamentous structures of cells by beginning with a cell population randomly spread on the flat surface of a homogeneous ECM [112].

One of the most important and well-known principles of developmental biology is the Differential Adhesion Hypothesis (DAH) proposed by Steinberg [113]. This hypothesis has gathered extensive experimental support and has now been widely accepted [114]. DAH states that (i) cell adhesion in multicellular systems is dependent on energy differences between different cell types and (ii) the cells or cell aggregates in a tissue are motile enough to reach the configuration where the energy is the lowest. DAH leads to a close analogy between living tissues (made of motile cells and adhesive) and true liquids [115].

A large number of discrete cell models are developed based on DAH. A typical model is Monte Carlo simulations (MCS) of the large-N Potts model, in which the tissue is represented on a lattice [116]. Each cell spans a few lattice sites and is given a unique identification number. The average number of sites per cell is maintained around a target value via an elastic energy term containing a Lagrange multiplier [109]. The simulations utilises the Metropolis algorithm, modelling cell migration and shape changes in systems consisting of up to several thousand cells [117]. The simulation results suggested that cell motility can be attributed to a temperature-like parameter.

In addition to the models based on DAH, computational models of *in vivo* morphogenesis recently developed include cell differentiation, chemical signalling, i.e. chemotaxis, and extracellular matrix production [118, 119]. The Glazier and Graner model [117] was combined with partial differential equations to describe cAMP signalling, by which the culmination process of the cellular slime mould was simulated in two dimensions. The parameter characteristic for the subcellular level was used to define the model, enabling it to predict phenomena where thousands of cells self-organise. The model has the potential to characterise the morphogenetic influence of genes whose function is explicated at the subcellular level [118]. Slime mould aggregation was described using a force-based 3D model where individual amoebae were treated as viscoelastic ellipsoids.

7.5.3 *Tissue spheroid and cell aggregate fusion*

In the evolving technology of bioprinting, successive layers of cell aggregates together with an embedding hydrogel are stacked on top of each other. Tissue liquidity is meant to lead to subsequent fusion of cell aggregates to obtain the constructs of desired shape. The *in silico* research of cellular rearrangement in post-processing stage may offer hints with regard to the conditions required to coax the cells to build the designed configuration [105, 120].

In long time scales, multicellular spheroids or aggregates of cells behave similarly to liquids. They tend to minimise their surface, and envelop or fuse each other while they are placed in contact [121, 122]. Cells in an aggregate interact in very complicated ways, which involves the following: (i) adhesion molecules that may attach to different types of cells, (ii) active motion resulting from the cytoskeleton activity, and (iii) the cortical tension initiating from the cytoskeleton structure that results in cells resisting to deformations. In terms of rheological properties, cell aggregates present elastic, viscous and plastic properties simultaneously [123]. Therefore, a detailed description of fusion of cell aggregates is necessary for yielding adequate control for bioprinting. The following subsections present some typical modelling approaches developed.

7.5.3.1 *A lattice model of living tissue*

A two-dimensional lattice model is a basic model mimicking a system of living cells in an extracellular matrix or a culture medium. This model then turns to simulate the evolution of cell aggregates using an algorithm (such as Monte Carlo algorithm to be introduced in the proceeding subsections) to develop a far more complicated three-dimensional model simulating tissue spheroid and cell aggregate fusion. Figure 7.5 depicts the 2D vision of a simplified square-lattice model. Each site is occupied either by a cell (green), or a similar-sized volume element of extracellular matrix or medium (grey). Every cell interacts to the same extent with the most adjacent or the next-most adjacent neighbours. Every cell interacts with neighbouring cells (highlighted by the dark lines) as well as with the surrounding medium (highlighted by the light lines).

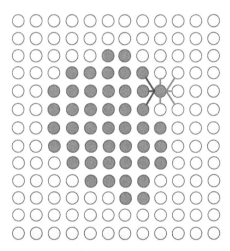

Fig. 7.5. A simplified 2D square-lattice model.

7.5.3.2 *Monte Carlo Simulations*

The Monte Carlo methods have been proven to be a convenient way to investigate conformational changes of a system driven by energy differences [124]. This name represents a collection of computational algorithms that use random numbers. The selection of random number generators is important and the selected generator should be able to generate pseudo-random numbers. In addition, the simulation results may be largely influenced by the correlations between the terms of the generated sequence. A popular random number generator is L'Ecuyer with Baym-Durham shuffle [125].

(i)   Metropolis Monte Carlo (MMC)

The Monte Carlo simulation can be further split into two methods, namely, Metropolis Monte Carlo (MMC) and kinetic Monte Carlo (KMC). For simulation of tissue evolution using MMC, the initial state is created and constructed based on the shape of a studied biological system and the known composition. A conformational change is made through identifying interfacial cells, selecting one of them by change and then exchanging it with a randomly chosen, adjacent medium particle. The corresponding adhesive energy change i.e. ΔE is calculated, and the new conformation is accepted with a probability P shown below.

$$P = \min\left(1, \exp\left(-\frac{\Delta E}{E_T}\right)\right) \qquad (7.1)$$

The move is accepted if it results in a decrease of the energy, whereas, in the opposite case, the move can also be considered as acceptable with a probability less than one (given by the Boltzmann factor). In fact, in order to make a decision, a random number, r, with uniform distribution between 0 and 1, is generated. The new conformation will be accepted as long as r is less than $\exp(-\Delta E/E_T)$. The sequence of operations, known as Monte Carlo step, is specified, during which each interfacial cell will be allowed to move once. A variation of MMC is cellular Potts model (CPM), which uses a modified MMC algorithm. The configuration of the simulated system is updated using the modified MMC algorithm and it is postulated that time is proportional to the number of Monte Carlo steps.

From the aspect of biology, the above stochastic rules indicate that a cell actively explores its neighbourhood. It can exchange its position with adjacent cells and the extracellular matrix can also be reorganised from their vicinity. The process of ECM reorganisation involves both enzymatic activities by matrix metalloproteases and mechanical traction forces. Such an algorithm has shown its advantages in studying fluids with velocity-independent interactions [126]. It provides a natural framework to explore the consequences of tissue liquidity.

(ii) Kinetic Monte Carlo

The Kinetic Monte Carlo method was developed as an alternative to the Metropolis Monte Carlo method for modelling the evolution of Ising models [127]. The main difference between KMC and MMC lies in the rejection rate. When a system is in a metastable state, or approaches equilibrium, the MMC method rejects most trial moves owing to the acceptance probability being small. In contrast, the KMC algorithm is 'rejection-free'. In each step, the KMC method calculates the transition rates for the changes compatible with the current configuration. Subsequently, a new configuration with a probability that is proportional to the corresponding transition rate is chosen. In other words, the KMC method can describe the transition rates in relation to the possible changes of the configuration of a multicellular system. These rates are then used to express the corresponding time evolution of the system. By doing so, a post-printed structure formation can be modelled where the multicellular system configuration is propagated in time. Simulation models based on KMC provide a more precise description of the time evolution for a multicellular system as compared with other grid-based methods such as MMC [128].

Figure 7.6 is an example of a KMC simulation with exterior and interior views of the spheroids, respectively. In the initial state, six uniluminal vascular spheroids are placed side by side along the x-axis (see Fig. 7.6(a) and (f)). The radius of each aggregate is approximately 7 cell diameters. There are 982 cells in each aggregate [129], which contains 680 vascular smooth muscle cells in the outer layer and 302 endothelial cells in the inner layer engulfing 436 lumen particles. Once the spheroids are in contact, partial fusion of the vascular smooth muscle cells in the outer layers ensues, as depicted in Fig. 7.6(b) and (g). After completing 2.5×10^6 steps, the lumens that are inside all the spheroids would have packed together and the fused tube shows a smooth outer surface (Fig. 7.6(e) and (j)). The biological analogues of this kind of structures are blood vessels.

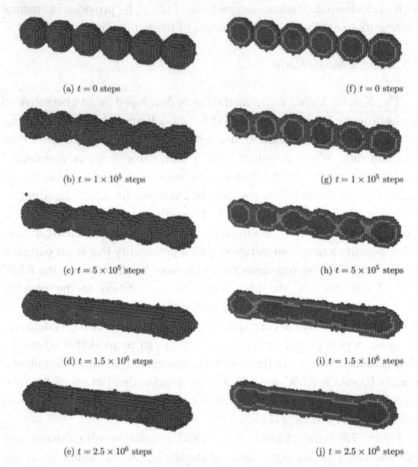

(a) $t = 0$ steps   (f) $t = 0$ steps

(b) $t = 1 \times 10^5$ steps   (g) $t = 1 \times 10^5$ steps

(c) $t = 5 \times 10^5$ steps   (h) $t = 5 \times 10^5$ steps

(d) $t = 1.5 \times 10^6$ steps   (i) $t = 1.5 \times 10^6$ steps

(e) $t = 2.5 \times 10^6$ steps   (j) $t = 2.5 \times 10^6$ steps

Fig. 7.6. Time evolution of tube formation via fusion of six uniluminal vascular spheroids in a KMC simulation [129] (Copyright 2013, with permission of Royal Society of Chemistry).

7.5.3.3 *Phase field approach*

The phase field method is proposed to study fusion of multicellular aggregates during tissue morphological development in bioprinting [130]. This method also aims to address the growing need to understand the mechanical-chemical properties of printed tissue constructs as well as the mechanism for morphological development during tissue fusion [131-133]. Multicellular aggregates or tissue spheroids are modelled as mixtures of multiphase complex fluids. The host hydrogel is treated as immiscible fluids where each type of cell forms its own fluid phase. The phase mixing or separation of these fluids is governed by interphase force interactions, imitating the cell-cell interaction in the tissue spheroids or aggregates. Additionally, the intermediate range interaction is mediated by the surrounding hydrogel [134]. This modelling method is applicable to transient simulations of cellular aggregate fusion at the length and time scale appropriate to bioprinting.

The simulation of the bioprinting of the vascular vein is shown in Fig. 7.7, where each ring/layer of spheroids is stacked on top of the other. There are three layers in the example. The initially rugged tube construct gradually evolves into a smooth tube, which qualitatively shows the morphological evolution of the spheroid fusion process in tube formation.

Fig. 7.7. Tube-formation of cellular spheroids in bioprinting. Snapshots are taken at different times [130] (Copyright 2012, with permission of Elsevier).

7.5.3.4 *Cellular particle dynamics (CPD)*

Another widely used simulation technique is cellular particle dynamics, which has been used to describe and predict the morphological evolution in time of three-dimensional multicellular systems during passive biomechanical relaxation [128]. This technique will be presented in the next subsection (modelling of printed structures).

7.5.4 *Modelling of printed structure*

CPD is an effective computational method to present the shape evolution and biomechanical relaxation processes in systems consisting of micro tissues e.g. multicellular aggregates. A representative relaxation process is the post bioprinting structure formation process where spheroidal bioink particles are fused together.

In CPD, cells are treated as a set of cellular particles (CPs) [135]. The total number of CPs is denoted as N_{CP}. Short-range contact forces drive interactions of CPs with both a repulsive (i.e. excluded volume) and an attractive (i.e. adhesive) component. Within a cell, CPs are gathered and held together by a confining potential. The potential maintains the cell integrity and functions as an effective cell membrane. The dynamics of CPs is presented by an over-damped Langevin equation [136]. The time evolution of the multicellular system conformation is determined by following the trajectories of all CPs through integrating their motion equations [135].

Quantitatively, the motion equation for the ith CP, $i \in [1, 2,..., N_{CP}]$, in cell n, $n \in [1, 2,...,N]$ is the Langevin equation [137] as depicted below.

$$\mu \frac{dr_{i,n}}{dt} = F_{1,i,n} + F_{2,i,n} + f_{i,n}(t) \qquad (7.2)$$

where, μ is the friction coefficient, $r_{i,n}$ is the position vector of the ith CP, $F_{1,i,n}$ is the force resulting from intracellular interactions exerted by CPs inside the nth cell, and $F_{2,i,n}$ is the force resulting from intracellular interactions exerted by CPs from other cells. $f_{i,n}(t)$ is a random force and is owing to the stochastic environment inside the cell. This environment is important in terms of cell motility as it can ultimately result in cell surface fluctuations [138]. $f_{i,n}(t)$ is modelled as a Gaussian white noise with zero mean and variance as shown below.

$$\langle f_{i,n,\alpha}(t) f_{j,m,\beta}(0) \rangle = 2D\mu^2 \delta(t) \delta_{ij} \delta_{nm} \delta_{\alpha\beta} \qquad (7.3)$$

where α and β denote vector component indices, and the variable D characterises the strength of the intracellular force fluctuations [135].

In addition, the parameters in the CPD model are determined by comparing the shape evolution of the experimental system and with that of the simulated system. The connection between the experimental and CPD simulation time units, is established by employing a theoretical continuum model. This model quantitatively describes the same multicellular fusion process that has previously been investigated both experimentally and through CPD simulations. Having established the connection, CPD simulations are able to quantitatively predict the evolution of the conformation of any multicellular system created by the same type of cell used in the experiments to identify and determine the CPD parameters [139].

References

[1] C. M. Cheah, C. K. Chua, K. F. Leong, C. H. Cheong, and M. W. Naing, "Automatic algorithm for generating complex polyhedral scaffold structures for tissue engineering," *Tissue Engineering*, vol. 10, pp. 595-610, 2004.

[2] F. W. Zonneveld, "3D imaging and its derivatives in clinical research and practice," *Critical Reviews in Computed Tomography*, vol. 41, pp. 69-156, 2000.

[3] I. Leichter, S. Fields, R. Nirel, P. Bamberger, B. Novak, R. Lederman, *et al.*, "Improved mammographic interpretation of masses using computer-aided diagnosis," *European radiology*, vol. 10, pp. 377-383, 2000.

[4] J. Y. Choi, J. H. Choi, N. K. Kim, Y. Kim, J. K. Lee, M. K. Kim, *et al.*, "Analysis of errors in medical rapid prototyping models," *International journal of oral and maxillofacial surgery*, vol. 31, pp. 23-32, 2002.

[5] R. Petzold, H. F. Zeilhofer, and W. Kalender, "Rapid prototyping technology in medicine—basics and applications," *Computerized Medical Imaging and Graphics*, vol. 23, pp. 277-284, 1999.

[6] K. Doi, H. MacMahon, S. Katsuragawa, R. M. Nishikawa, and Y. Jiang, "Computer-aided diagnosis in radiology: potential and pitfalls," *European journal of Radiology*, vol. 31, pp. 97-109, 1999.

[7] S. Giannitelli, D. Accoto, M. Trombetta, and A. Rainer, "Current trends in the design of scaffolds for computer-aided tissue engineering," *Acta biomaterialia,* vol. 10, pp. 580-594, 2014.

[8] W. Sun, A. Darling, B. Starly, and J. Nam, "Computer-aided tissue engineering: overview, scope and challenges," *Biotechnology and Applied Biochemistry,* vol. 39, pp. 29-47, 2004.

[9] W. Sun and P. Lal, "Recent development on computer aided tissue engineering—a review," *Computer methods and programs in biomedicine,* vol. 67, pp. 85-103, 2002.

[10] R. Landers, U. Hübner, R. Schmelzeisen, and R. Mülhaupt, "Rapid prototyping of scaffolds derived from thermoreversible hydrogels and tailored for applications in tissue engineering," *Biomaterials,* vol. 23, pp. 4437-4447, 2002.

[11] B. Van Rietbergen, R. Müller, D. Ulrich, P. Rüegsegger, and R. Huiskes, "Tissue stresses and strain in trabeculae of a canine proximal femur can be quantified from computer reconstructions," *Journal of biomechanics,* vol. 32, pp. 165-173, 1999.

[12] R. Müller and P. Rüegsegger, "Micro-tomographic imaging for the nondestructive evaluation of trabecular bone architecture," *Studies in Health Technology and Informatics,* pp. 61-80, 1997.

[13] A. K. A. Breithecker and W. Rau, "3D imaging of lung tissue by confocal microscopy and micro-CT," in *Laser-tissue Interaction,* 2001, pp. 469-476.

[14] R. Müller, S. Matter, P. Neuenschwander, U. Suter, and P. Rüegsegger, "3D micro-tomographic imaging and quantitative morphometry for the nondestructive evaluation of porous biomaterials," in *MRS Proceedings,* 1996, p. 217.

[15] W. Krause, K. Handreke, G. Schuhmann-Giampieri, and K. Rupp, "Efficacy of the iodine-free computed tomography liver contrast agent, Dy-EOB-DTPA, in comparison with a conventional iodinated agent in normal and in tumor-bearing rabbits," *Investigative radiology,* vol. 37, pp. 241-247, 2002.

[16] M. Watanabe, T. Shin'oka, S. Tohyama, N. Hibino, T. Konuma, G. Matsumura, et al., "Tissue-engineered vascular autograft: inferior vena cava replacement in a dog model," *Tissue engineering,* vol. 7, pp. 429-439, 2001.

[17] M. Dhenain, S. W. Ruffins, and R. E. Jacobs, "Three-dimensional digital mouse atlas using high-resolution MRI," *Developmental biology,* vol. 232, pp. 458-470, 2001.

[18] R. Karch, F. Neumann, M. Neumann, and W. Schreiner, "Staged growth of optimized arterial model trees," *Annals of biomedical engineering,* vol. 28, pp. 495-511, 2000.

[19] K. H. Hohne, "Medical image computing at the institute of mathematics and computer science in medicine, university hospital hamburg-eppendorf," *Medical Imaging, IEEE Transactions on,* vol. 21, pp. 713-723, 2002.

[20] N. J. Mankovich, D. R. Robertson, and A. M. Cheeseman, "Three-dimensional image display in medicine," *Journal of digital imaging,* vol. 3, pp. 69-80, 1990.

[21] M. Viceconti, C. Zannoni, and L. Pierotti, "TRI2SOLID: an application of reverse engineering methods to the creation of CAD models of bone segments," *Computer methods and programs in biomedicine,* vol. 56, pp. 211-220, 1998.

[22] M. Viceconti, C. Zannoni, D. Testi, and A. Cappello, "CT data sets surface extraction for biomechanical modeling of long bones," *Computer methods and programs in biomedicine,* vol. 59, pp. 159-166, 1999.

[23] P. Sabella, "A rendering algorithm for visualizing 3D scalar fields," in *ACM SIGGRAPH computer graphics,* 1988, pp. 51-58.

[24] K. Mueller and R. Yagel, "Fast perspective volume rendering with splatting by utilizing a ray-driven approach," in *Proceedings of the 7th conference on Visualization'96,* 1996, pp. 65-72.

[25] J. Freund and K. Sloan, "Accelerated volume rendering using homogeneous region encoding," in *Proceedings of the 8th conference on Visualization'97,* 1997, pp. 191-ff.

[26] S. M. Morvan and G. M. Fadel, "Heterogeneous solids: possible representation schemes," in *Proceedings of the Solid Freeform Fabrication Symposium,* Austin, Taxas, USA, 1999, pp. 187-197.

[27] N. J. Mankovich, A. M. Cheeseman, and N. G. Stoker, "The display of three-dimensional anatomy with stereolithographic models," *Journal of digital imaging,* vol. 3, pp. 200-203, 1990.

[28] F. Pazos and A. Valencia, "In silico two-hybrid system for the selection of physically interacting protein pairs," *Proteins:*

Structure, Function, and Bioinformatics, vol. 47, pp. 219-227, 2002.

[29]   A. B. Olshen and A. N. Jain, "Deriving quantitative conclusions from microarray expression data," *Bioinformatics,* vol. 18, pp. 961-970, 2002.

[30]   J. Angulo and J. Serra, "Automatic analysis of DNA microarray images using mathematical morphology," *Bioinformatics,* vol. 19, pp. 553-562, 2003.

[31]   A. Wee Chung Liew, H. Yan, and M. Yang, "Robust adaptive spot segmentation of DNA microarray images," *Pattern Recognition,* vol. 36, pp. 1251-1254, 2003.

[32]   D. Meier and E. Fisher, "Parameter space warping: Shape-based correspondence between morphologically different objects," *Medical Imaging, IEEE Transactions on,* vol. 21, pp. 31-47, 2002.

[33]   C. K. Chua, N. Sudarmadji, and K. F. Leong, "Functionally Graded Scaffolds: the Challenges in Design and Fabrication Methods," in *3rd International Conference on Advanced Research in Virtual and Rapid Prototyping,* Leiria, Portugal, 2007.

[34]   K. F. Leong, C. K. Chua, N. Sudarmadji, and W. Y. Yeong, "Engineering functionally graded tissue engineering scaffolds," *Journal of the mechanical behavior of biomedical materials,* vol. 1, pp. 140-152, 2008.

[35]   D. W. Hutmacher, "Scaffolds in tissue engineering bone and cartilage," *Biomaterials,* vol. 21, pp. 2529-2543, 2000.

[36]   S. Hollister, R. Maddox, and J. Taboas, "Optimal design and fabrication of scaffolds to mimic tissue properties and satisfy biological constraints," *Biomaterials,* vol. 23, pp. 4095-4103, 2002.

[37]   B. Starly, C. Gomez, A. Darling, Z. Fang, A. Lau, W. Sun, *et al.,* "Computer-aided bone scaffold design: a biomimetic approach," in *2003 IEEE 29th Annual Bioengineering Conference,* Newark, New Jersey, USA, 2003, pp. 172-173.

[38]   B. Starly, A. Darling, and W. Sun, "Biomimetic design of 3D bone scaffolds," in *Proceedings of 2002 meeting on Tissue Engineering at Cold Spring Harbor Laboratory, Cold Spring Harbor, New York,* 2002.

[39]   W. Sun, B. Starly, A. Darling, and C. Gomez, "Computer-aided tissue engineering: application to biomimetic modelling and

design of tissue scaffolds," *Biotechnology and applied biochemistry,* vol. 39, pp. 49-58, 2004.

[40] W. Sun, B. Starly, J. Nam, and A. Darling, "Bio-CAD modeling and its applications in computer-aided tissue engineering," *Computer-Aided Design,* vol. 37, pp. 1097-1114, 2005.

[41] Z. Fang, B. Starly, and W. Sun, "Computer-aided characterization for effective mechanical properties of porous tissue scaffolds," *Computer-Aided Design,* vol. 37, pp. 65-72, 2005.

[42] C. M. Cheah, C. K. Chua, K. F. Leong, and S. W. Chua, "Development of a tissue engineering scaffold structure library for rapid prototyping. Part 1: investigation and classification," *The International Journal of Advanced Manufacturing Technology,* vol. 21, pp. 291-301, 2003.

[43] C. M. Cheah, C. K. Chua, K. F. Leong, and S. W. Chua, "Development of a tissue engineering scaffold structure library for rapid prototyping. Part 2: parametric library and assembly program," *The International Journal of Advanced Manufacturing Technology,* vol. 21, pp. 302-312, 2003.

[44] C. K. Chua, M. W. Naing, K. F. Leong, and C. M. Cheah, "Novel Method for Producing Polyhedral Scaffolds in Tissue Engineering," in *International Conference on Advanced Research in Virtual and Rapid Prototyping,* Leiria, Portugal, 2003.

[45] C. K. Chua, K. F. Leong, N. Sudarmadji, M. J. J. Liu, and S. M. Chou, "Selective laser sintering of functionally graded tissue scaffolds," *MRS bulletin,* vol. 36, pp. 1006-1014, 2011.

[46] N. Sudarmadji, J. Y. Tan, K. F. Leong, C. K. Chua, and Y. T. Loh, "Investigation of the mechanical properties and porosity relationships in selective laser-sintered polyhedral for functionally graded scaffolds," *Acta biomaterialia,* vol. 7, pp. 530-537, 2011.

[47] W. Sun and X. Hu, "Reasoning Boolean operation based modeling for heterogeneous objects," *Computer-Aided Design,* vol. 34, pp. 481-488, 2002.

[48] S. J. Hollister, R. A. Levy, T. M. Chu, J. W. Halloran, and S. E. Feinberg, "An image-based approach for designing and manufacturing craniofacial scaffolds," *International Journal of Oral & Maxillofacial Surgery,* vol. 29, pp. 67-71, 2000.

[49]   S. E. Feinberg, S. J. Hollister, J. W. Halloran, T. G. Chu, and P. H. Krebsbach, "Image-based biomimetic approach to reconstruction of the temporomandibular joint," *Cells Tissues Organs,* vol. 169, pp. 309-321, 2001.

[50]   M. Smith, C. Flanagan, J. Kemppainen, J. Sack, H. Chung, S. Das, *et al.,* "Computed tomography-based tissue-engineered scaffolds in craniomaxillofacial surgery," *The International Journal of Medical Robotics and Computer Assisted Surgery,* vol. 3, pp. 207-216, 2007.

[51]   S. J. Hollister, "Porous scaffold design for tissue engineering," *Nature materials,* vol. 4, pp. 518-524, 2005.

[52]   W. Chiu, Y. Yeung, and K. Yu, "Toolpath generation for layer manufacturing of fractal objects," *Rapid Prototyping Journal,* vol. 12, pp. 214-221, 2006.

[53]   K. W. Lee, S. Wang, M. Dadsetan, M. J. Yaszemski, and L. Lu, "Enhanced cell ingrowth and proliferation through three-dimensional nanocomposite scaffolds with controlled pore structures," *Biomacromolecules,* vol. 11, pp. 682-689, 2010.

[54]   B. Duan, W. L. Cheung, and M. Wang, "Optimized fabrication of Ca–P/PHBV nanocomposite scaffolds via selective laser sintering for bone tissue engineering," *Biofabrication,* vol. 3, p. 015001, 2011.

[55]   N. Chantarapanich, P. Puttawibul, S. Sucharitpwatskul, P. Jeamwatthanachai, S. Inglam, and K. Sitthiseripratip, "Scaffold library for tissue engineering: a geometric evaluation," *Computational and mathematical methods in medicine,* vol. 2012, 2012.

[56]   M. Wettergreen, B. Bucklen, B. Starly, E. Yuksel, W. Sun, and M. Liebschner, "Creation of a unit block library of architectures for use in assembled scaffold engineering," *Computer-Aided Design,* vol. 37, pp. 1141-1149, 2005.

[57]   R. Gabbrielli, I. Turner, and C. R. Bowen, "Development of modelling methods for materials to be used as bone substitutes," *Key Engineering Materials,* vol. 361, pp. 903-906, 2008.

[58]   A. Pasko, O. Fryazinov, T. Vilbrandt, P.-A. Fayolle, and V. Adzhiev, "Procedural function-based modelling of volumetric microstructures," *Graphical Models,* vol. 73, pp. 165-181, 2011.

[59]   M. E. Hoque, Y. L. Chuan, and I. Pashby, "Extrusion based rapid prototyping technique: An advanced platform for tissue

engineering scaffold fabrication," *Biopolymers,* vol. 97, pp. 83-93, 2012.

[60] I. Zein, D. W. Hutmacher, K. C. Tan, and S. H. Teoh, "Fused deposition modeling of novel scaffold architectures for tissue engineering applications," *Biomaterials,* vol. 23, pp. 1169-1185, 2002.

[61] T. B. Woodfield, L. Moroni, and J. Malda, "Combinatorial approaches to controlling cell behaviour and tissue formation in 3D via rapid-prototyping and smart scaffold design," *Combinatorial chemistry & high throughput screening,* vol. 12, pp. 562-579, 2009.

[62] J. Li, J. De Wijn, C. Van Blitterswijk, and K. De Groot, "The effect of scaffold architecture on properties of direct 3D fiber deposition of porous Ti6Al4V for orthopedic implants," *Journal of Biomedical Materials Research Part A,* vol. 92, pp. 33-42, 2010.

[63] H. Sagan, *Space-filling curves* vol. 18. Berlin: Springer-Verlag, 1994.

[64] B. Starly and W. Sun, "Internal scaffold architecture designs using lindenmayer systems," *Computer-Aided Design and Applications,* vol. 4, pp. 395-403, 2007.

[65] G. S. Kumar, P. Pandithevan, and A. R. Ambatti, "Fractal raster tool paths for layered manufacturing of porous objects," *Virtual and Physical Prototyping,* vol. 4, pp. 91-104, 2009.

[66] X. Kou and S. Tan, "A simple and effective geometric representation for irregular porous structure modeling," *Computer-Aided Design,* vol. 42, pp. 930-941, 2010.

[67] D. W. Hutmacher, M. Sittinger, and M. V. Risbud, "Scaffold-based tissue engineering: rationale for computer-aided design and solid free-form fabrication systems," *Trends in Biotechnology,* vol. 22, pp. 354-362, 2004.

[68] Z. Chen, Z. Su, S. Ma, X. Wu, and Z. Luo, "Biomimetic modeling and three-dimension reconstruction of the artificial bone," *Computer methods and programs in biomedicine,* vol. 88, pp. 123-130, 2007.

[69] S. J. Kalita, S. Bose, H. L. Hosick, and A. Bandyopadhyay, "Development of controlled porosity polymer-ceramic composite scaffolds via fused deposition modeling," *Materials Science and Engineering: C,* vol. 23, pp. 611-620, 2003.

[70]   A. Khoda, I. T. Ozbolat, and B. Koc, "Engineered tissue scaffolds with variational porous architecture," *Journal of biomechanical engineering*, vol. 133, p. 011001, 2011.

[71]   C. Schroeder, W. C. Regli, A. Shokoufandeh, and W. Sun, "Computer-aided design of porous artifacts," *Computer-Aided Design*, vol. 37, pp. 339-353, 2005.

[72]   S. Sogutlu and B. Koc, "Stochastic modeling of tissue engineering scaffolds with varying porosity levels," *Computer-Aided Design and Applications*, vol. 4, pp. 661-670, 2007.

[73]   D. Lacroix, J. A. Planell, and P. J. Prendergast, "Computer-aided design and finite-element modelling of biomaterial scaffolds for bone tissue engineering," *Philosophical Transactions of the Royal Society A: Mathematical, Physical and Engineering Sciences*, vol. 367, pp. 1993-2009, 2009.

[74]   S. Cahill, S. Lohfeld, and P. McHugh, "Finite element predictions compared to experimental results for the effective modulus of bone tissue engineering scaffolds fabricated by selective laser sintering," *Journal of Materials Science: Materials in Medicine*, vol. 20, pp. 1255-1262, 2009.

[75]   G. Ryan, P. McGarry, A. Pandit, and D. Apatsidis, "Analysis of the mechanical behavior of a titanium scaffold with a repeating unit-cell substructure," *Journal of Biomedical Materials Research Part B: Applied Biomaterials*, vol. 90, pp. 894-906, 2009.

[76]   E. Saito, H. Kang, J. M. Taboas, A. Diggs, C. L. Flanagan, and S. J. Hollister, "Experimental and computational characterization of designed and fabricated 50: 50 PLGA porous scaffolds for human trabecular bone applications," *Journal of Materials Science: Materials in Medicine*, vol. 21, pp. 2371-2383, 2010.

[77]   J. M. Williams, A. Adewunmi, R. M. Schek, C. L. Flanagan, P. H. Krebsbach, S. E. Feinberg, *et al.*, "Bone tissue engineering using polycaprolactone scaffolds fabricated via selective laser sintering," *Biomaterials*, vol. 26, pp. 4817-4827, 2005.

[78]   S. Eshraghi and S. Das, "Mechanical and microstructural properties of polycaprolactone scaffolds with one-dimensional, two-dimensional, and three-dimensional orthogonally oriented porous architectures produced by selective laser sintering," *Acta Biomaterialia*, vol. 6, pp. 2467-2476, 2010.

[79]   S. Eshraghi and S. Das, "Micromechanical finite-element modeling and experimental characterization of the compressive

mechanical properties of polycaprolactone–hydroxyapatite composite scaffolds prepared by selective laser sintering for bone tissue engineering," *Acta biomaterialia,* vol. 8, pp. 3138-3143, 2012.

[80] J. W. Jung, H. G. Yi, T. Y. Kang, W. J. Yong, S. Jin, W. S. Yun, *et al.,* "Evaluation of the effective diffusivity of a freeform fabricated scaffold using computational simulation," *Journal of biomechanical engineering,* vol. 135, p. 084501, 2013.

[81] C. X. Lam, M. M. Savalani, S. H. Teoh, and D. W. Hutmacher, "Dynamics of in vitro polymer degradation of polycaprolactone-based scaffolds: accelerated versus simulated physiological conditions," *Biomedical Materials,* vol. 3, p. 034108, 2008.

[82] E. Saito, Y. Liu, F. Migneco, and S. J. Hollister, "Strut size and surface area effects on long-term in vivo degradation in computer designed poly (l-lactic acid) three-dimensional porous scaffolds," *Acta biomaterialia,* vol. 8, pp. 2568-2577, 2012.

[83] J. Malda, T. B. Woodfield, F. Van Der Vloodt, F. Kooy, D. E. Martens, J. Tramper, *et al.,* "The effect of PEGT/PBT scaffold architecture on oxygen gradients in tissue engineered cartilaginous constructs," *Biomaterials,* vol. 25, pp. 5773-5780, 2004.

[84] S. Truscello, G. Kerckhofs, S. Van Bael, G. Pyka, J. Schrooten, and H. Van Oosterwyck, "Prediction of permeability of regular scaffolds for skeletal tissue engineering: a combined computational and experimental study," *Acta biomaterialia,* vol. 8, pp. 1648-1658, 2012.

[85] A. L. Olivares, È. Marsal, J. A. Planell, and D. Lacroix, "Finite element study of scaffold architecture design and culture conditions for tissue engineering," *Biomaterials,* vol. 30, pp. 6142-6149, 2009.

[86] H. Singh, S. H. Teoh, H. T. Low, and D. Hutmacher, "Flow modelling within a scaffold under the influence of uni-axial and bi-axial bioreactor rotation," *Journal of biotechnology,* vol. 119, pp. 181-196, 2005.

[87] D. W. Hutmacher and H. Singh, "Computational fluid dynamics for improved bioreactor design and 3D culture," *Trends in biotechnology,* vol. 26, pp. 166-172, 2008.

[88] B. Porter, R. Zauel, H. Stockman, R. Guldberg, and D. Fyhrie, "3-D computational modeling of media flow through scaffolds in

a perfusion bioreactor," *Journal of biomechanics,* vol. 38, pp. 543-549, 2005.

[89]   Y. Yao, W. D. Chen, and W. Y. Jin, "The influence of pore structure on internal flow field shear stress within scaffold," *Advanced Materials Research,* vol. 308, pp. 771-775, 2011.

[90]   M. P. Bendsoe and O. Sigmund, *Topology optimization: theory, methods and applications.* London: Springer-Verlag, 2003.

[91]   C. Y. Lin, N. Kikuchi, and S. J. Hollister, "A novel method for biomaterial scaffold internal architecture design to match bone elastic properties with desired porosity," *Journal of biomechanics,* vol. 37, pp. 623-636, 2004.

[92]   V. J. Challis, A. P. Roberts, J. F. Grotowski, L. C. Zhang, and T. B. Sercombe, "Prototypes for bone implant scaffolds designed via topology optimization and manufactured by solid freeform fabrication," *Advanced Engineering Materials,* vol. 12, pp. 1106-1110, 2010.

[93]   M. W. Naing, C. K. Chua, K. F. Leong, and Y. Wang, "Fabrication of customised scaffolds using computer-aided design and rapid prototyping techniques," *Rapid Prototyping Journal,* vol. 11, pp. 249-259, 2005.

[94]   N. Sudarmadji, C. K. Chua, and K. F. Leong, "The Development of Computer-Aided System for Tissue Scaffolds (CASTS) System for Functionally Graded Tissue-Engineering Scaffolds," in *Computer-Aided Tissue Engineering,* ed: Springer, 2012, pp. 111-123.

[95]   C. Laurent, D. Durville, C. Vaquette, R. Rahouadj, and J.-F. Ganghoffer, "Computer-Aided Tissue Engineering: Application to the Case of Anterior Cruciate Ligament Repair," in *Biomechanics of Cells and Tissues,* ed: Springer, 2013, pp. 1-44.

[96]   C. P. Laurent, D. Durville, D. Mainard, J.-F. Ganghoffer, and R. Rahouadj, "A multilayer braided scaffold for Anterior Cruciate Ligament: Mechanical modeling at the fiber scale," *Journal of the mechanical behavior of biomedical materials,* vol. 12, pp. 184-196, 2012.

[97]   F. C. Fierz, F. Beckmann, M. Huser, S. H. Irsen, B. Leukers, F. Witte, *et al.,* "The morphology of anisotropic 3D-printed hydroxyapatite scaffolds," *Biomaterials,* vol. 29, pp. 3799-3806, 2008.

[98] V. Karageorgiou and D. Kaplan, "Porosity of 3D biomaterial scaffolds and osteogenesis," *Biomaterials,* vol. 26, pp. 5474-5491, 2005.

[99] A. J. Stops, K. Heraty, M. Browne, F. J. O'Brien, and P. McHugh, "A prediction of cell differentiation and proliferation within a collagen–glycosaminoglycan scaffold subjected to mechanical strain and perfusive fluid flow," *Journal of biomechanics,* vol. 43, pp. 618-626, 2010.

[100] K. Y. Volokh, "Stresses in growing soft tissues," *Acta Biomaterialia,* vol. 2, pp. 493-504, Sep 2006.

[101] E. Kuhl and P. Steinmann, "Mass–and volume–specific views on thermodynamics for open systems," *Proceedings of the Royal Society of London. Series A: Mathematical, Physical and Engineering Sciences,* vol. 459, pp. 2547-2568, 2003.

[102] K. Garikipati, E. Arruda, K. Grosh, H. Narayanan, and S. Calve, "A continuum treatment of growth in biological tissue: the coupling of mass transport and mechanics," *Journal of the Mechanics and Physics of Solids,* vol. 52, pp. 1595-1625, 2004.

[103] A. Menzel, "Modelling of anisotropic growth in biological tissues," *Biomechanics and modeling in mechanobiology,* vol. 3, pp. 147-171, 2005.

[104] S. M. Klisch and A. Hoger, "Volumetric growth of thermoelastic materials and mixtures," *Mathematics and Mechanics of Solids,* vol. 8, pp. 377-402, 2003.

[105] J. M. Pérez Pomares and R. A. Foty, "Tissue fusion and cell sorting in embryonic development and disease: biomedical implications," *Bioessays,* vol. 28, pp. 809-821, 2006.

[106] V. Mironov, V. Kasyanov, C. Drake, and R. R. Markwald, "Organ printing: promises and challenges," 2008.

[107] K. Jakab, A. Neagu, V. Mironov, R. R. Markwald, and G. Forgacs, "Engineering biological structures of prescribed shape using self-assembling multicellular systems," *Proceedings of the National Academy of Sciences of the United States of America,* vol. 101, pp. 2864-2869, 2004.

[108] K. Jakab, B. Damon, A. Neagu, A. Kachurin, and G. Forgacs, "Three-dimensional tissue constructs built by bioprinting," *Biorheology,* vol. 43, pp. 509-513, 2006.

[109] A. Neagu, I. Kosztin, K. Jakab, B. Barz, M. Neagu, R. Jamison, *et al.,* "Computational modeling of tissue self-assembly," *Modern Physics Letters B,* vol. 20, pp. 1217-1231, Aug 2006.

[110]   J. L. Semple, N. Woolridge, and C. J. Lumsden, "Review: in vitro, in vivo, in silico: computational systems in tissue engineering and regenerative medicine," *Tissue engineering,* vol. 11, pp. 341-356, 2005.

[111]   J. Murray, D. Manoussaki, S. Lubkin, and R. Vernon, "A mechanical theory of in vitro vascular network formation," in *Vascular morphogenesis: in vivo, in vitro, in mente,* ed: Springer, 1996, pp. 173-188.

[112]   D. Manoussaki, S. Lubkin, R. Vemon, and J. Murray, "A mechanical model for the formation of vascular networks in vitro," *Acta Biotheoretica,* vol. 44, pp. 271-282, 1996.

[113]   M. S. Steinberg, "Reconstruction of tissues by dissociated cells," *Science,* vol. 141, pp. 401-408, 1963.

[114]   D. Gonzalez Rodriguez, K. Guevorkian, S. Douezan, and F. Brochard Wyart, "Soft matter models of developing tissues and tumors," *Science,* vol. 338, pp. 910-917, 2012.

[115]   G. Forgacs and S. A. Newman, *Biological physics of the developing embryo.* New York: Cambridge University Press, 2005.

[116]   F. Graner and J. A. Glazier, "Simulation of biological cell sorting using a two-dimensional extended Potts model," *Physical review letters,* vol. 69, p. 2013, 1992.

[117]   J. A. Glazier and F. Graner, "Simulation of the differential adhesion driven rearrangement of biological cells," *Physical Review E,* vol. 47, p. 2128, 1993.

[118]   A. F. Marée and P. Hogeweg, "How amoeboids self-organize into a fruiting body: multicellular coordination in Dictyostelium discoideum," *Proceedings of the National Academy of Sciences,* vol. 98, pp. 3879-3883, 2001.

[119]   E. Palsson and H. G. Othmer, "A model for individual and collective cell movement in Dictyostelium discoideum," *Proceedings of the National Academy of Sciences,* vol. 97, pp. 10448-10453, 2000.

[120]   A. Wessels and D. Sedmera, "Developmental anatomy of the heart: a tale of mice and man," *Physiological genomics,* vol. 15, pp. 165-165, 2004.

[121]   R. Z. Lin and H. Y. Chang, "Recent advances in three-dimensional multicellular spheroid culture for biomedical research," *Biotechnology Journal,* vol. 3, pp. 1172-1184, 2008.

[122] J. Mombach, D. Robert, F. Graner, G. Gillet, G. L. Thomas, M. Idiart, *et al.*, "Rounding of aggregates of biological cells: Experiments and simulations," *Physica A: Statistical Mechanics and its Applications,* vol. 352, pp. 525-534, 2005.

[123] P. Marmottant, A. Mgharbel, J. Käfer, B. Audren, J.-P. Rieu, J.-C. Vial, *et al.*, "The role of fluctuations and stress on the effective viscosity of cell aggregates," *Proceedings of the National Academy of Sciences,* vol. 106, pp. 17271-17275, 2009.

[124] J. G. Amar, "The Monte Carlo method in science and engineering," *Computing in Science & Engineering,* vol. 8, pp. 9-19, 2006.

[125] W. H. Press, *Numerical recipes 3rd edition: The art of scientific computing.* New York: Cambridge University Press, Cambridge, UK, 2007.

[126] Y. Shim and J. G. Amar, "Hybrid asynchronous algorithm for parallel kinetic Monte Carlo simulations of thin film growth," *Journal of Computational Physics,* vol. 212, pp. 305-317, 2006.

[127] A. B. Bortz, M. H. Kalos, and J. L. Lebowitz, "A new algorithm for Monte Carlo simulation of Ising spin systems," *Journal of Computational Physics,* vol. 17, pp. 10-18, 1975.

[128] E. Flenner, L. Janosi, B. Barz, A. Neagu, G. Forgacs, and I. Kosztin, "Kinetic Monte Carlo and cellular particle dynamics simulations of multicellular systems," *Physical Review E,* vol. 85, Mar 2012.

[129] Y. Sun and Q. Wang, "Modeling and simulations of multicellular aggregate self-assembly in biofabrication using kinetic Monte Carlo methods," *Soft Matter,* vol. 9, pp. 2172-2186, 2013.

[130] X. F. Yang, V. Mironov, and Q. Wang, "Modeling fusion of cellular aggregates in biofabrication using phase field theories," *Journal of Theoretical Biology,* vol. 303, pp. 110-118, Jun 2012.

[131] K. Jakab, C. Norotte, B. Damon, F. Marga, A. Neagu, C. L. Besch-Williford, *et al.*, "Tissue engineering by self-assembly of cells printed into topologically defined structures," *Tissue Engineering Part A,* vol. 14, pp. 413-421, 2008.

[132] A. N. Mehesz, J. Brown, Z. Hajdu, W. Beaver, J. da Silva, R. Visconti, *et al.*, "Scalable robotic biofabrication of tissue spheroids," *Biofabrication,* vol. 3, p. 025002, 2011.

[133] K. Jakab, C. Norotte, F. Marga, K. Murphy, G. Vunjak-Novakovic, and G. Forgacs, "Tissue engineering by self-

assembly and bio-printing of living cells," *Biofabrication,* vol. 2, p. 022001, 2010.

[134] X. F. Yang, Y. Sun, and Q. Wang, "A Phase Field Approach for Multicellular Aggregate Fusion in Biofabrication," *Journal of Biomechanical Engineering-Transactions of the Asme,* vol. 135, Jul 2013.

[135] E. Flenner, L. Janosi, B. Barz, A. Neagu, G. Forgacs, and I. Kosztin, "Kinetic Monte Carlo and cellular particle dynamics simulations of multicellular systems," *Physical Review E,* vol. 85, p. 031907, 2012.

[136] E. Flenner, F. Marga, A. Neagu, I. Kosztin, and G. Forgacs, "Relating biophysical properties across scales," *Current topics in developmental biology,* vol. 81, pp. 461-483, 2008.

[137] T. Lubensky and P. Chaikin, "Principles of condensed matter physics," ed. New York: Cambridge University Press, Cambridge, UK, 1995.

[138] R. Gordon, N. S. Goel, M. S. Steinberg, and L. L. Wiseman, "A rheological mechanism sufficient to explain the kinetics of cell sorting," *Journal of theoretical biology,* vol. 37, pp. 43-73, 1972.

[139] M. McCune, A. Shafiee, G. Forgacs, and I. Kosztin, "Predictive modeling of post bioprinting structure formation," *Soft Matter,* vol. 10, pp. 1790-1800, 2014.

Problems

1. What are the three imaging modalities that have been widely used in tissue modelling? Describe their working principles and compare their advantages and disadvantages.
2. Describe the major steps in the manufacture of a patient model by using additive manufacturing (AM).
3. Describe the biological, mechanical and anatomical requirements for biomimetic modelling in tissue scaffold designs.
4. Describe two popular architectures used in tissue scaffold designs.
5. What are the strategies that can be used to optimise tissue scaffold designs?
6. What are the 4 major techniques for modelling tissue spheroid and cell aggregate fusion? Describe the modelling principles and compare their advantages and disadvantages.

Chapter 8

Applications of Bioprinting: Challenges and Potential

Direct printing of living functional organs is the ultimate goal of the three-dimensional (3D) bioprinting technology. If this becomes true eventually, it will be a revolutionary technology for medical industry. As this technology is still at the initial stage of development at present, it is early and unrealistic to speculate when these printed organs will become available. Kidneys, livers and hearts are commonly believed to be the organs that bioprinting should aim for in the future. Based on the waiting lists of kidney, liver and heart transplants in the USA and Europe, the bioprinted organ market is expected to increase with an annual growth of over 100% [1].

The bioprinting research and development will drive the growth of bioprinting market, which has also shown the potential to replace existing markets of other TE products. 3D bioprinting has already been used for the generation and transplantation of several tissues, including skin, bone, vascular grafts, heart tissue and cartilage tissue. Other applications include developing high-throughput 3D-bioprinted tissue models for research, drug discovery and toxicology [2]. As a nascent and emerging field, there are several challenges that need to be overcome for successful implementation of 3D bioprinting, especially in light of regulatory, technical and operational issues.

8.1   Challenges of Bioprinting

8.1.1   *Regulatory issues*

The regulatory constraints are one of the most challenging barriers for introducing bioprinting as a mainstream technology for drug companies. Based on the current Food and Drug Administration (FDA) categories, bioprinting does not fit in any one of them. FDA has still been considering new and far more restrictive regulations to be put in place specifically for the safe use of bioprinters and bioprinted products.

In addition, regulatory authorities including FDA currently require drug developers to carry out preclinical research on animals. Although bioprinted human micro-tissues will be used for internal research before the pharmaceutical companies take the molecules to preclinical, it still requires changing the existing policies for bioprinted micro-tissues to fully replace animal testing.

8.1.2   *Technical and operational issues*

8.1.2.1   *Technical challenges*

The major technical challenges lie in the following four aspects:

(i)   Design of construct

A computer-aided design (CAD) file is necessary to initiate the layer-by-layer bioprinting process. The construct of interest is of dynamic nature, involving tissue fusion, compaction and retraction and constant remodelling. Thus the initial design file might not resemble the final intended construct for implantation. It is also critical to include compensation factor in the initial design to compensate for the remodelling events.

(ii) Bio-ink development

The bio-ink or material selected for bioprinting needs to fulfil several requirements in terms of functional, mechanical, printability and biocompatibility with the printed cells. The material must also be capable of supporting cell-specific events such as tissue fusion, cell proliferation and migration. The bio-ink should completely degrade away while the cell population expands and matures into a biological construct.

(iii) Bioprinting technology

The spatial distribution of human tissue spans across a spectrum of length scale from nm to mm. It is challenging to replicate the full range of resolution via a specific bioprinting technique.

(iv) Integration with host tissue via vascularisation

The printed construct will be implanted and thus needs to be integrated with the host tissue for normal tissue function. The integration requires sufficient connection of vascularisation and innervation. Vascular networks supply oxygen and necessary nutrients to the organs. Blood vessels have been successfully printed together with endothelial cell aggregates, cartilage and skin tissues. However, printing vascular tubes into more complex tissues still remains unsolved. The bioprinting technology is currently unable to create very small capillaries linking larger vessels to cells.

8.1.2.2 *Operational challenges*

The two major operational challenges are outlined below:

(i) Cost of bioprinting technology

The price, as high as USD 200,000 for an Organovo's or EnvisionTEC's bioprinter, is a major limitation to spreading the bioprinting technology widely. Even though this revolutionary technology is expected to significantly reduce the costs involved in drug discovery, the costly initial investment restricts the current application of the technology.

(ii) Development of enabling technologies

The implementation of bioprinting technology will require several enabling technologies such as a scalable method for tissue spheroid fabrication, bioreactor technology for maturation of bioprinted construct, computational design capability for designing complex organ and advancement of cell processing technologies to obtain reproducible source of cells.

8.2 Potential of 3D Bioprinting

8.2.1 *Skin Tissue Engineering*

One of the current popular treatments for severely burnt patients is skin grafting. Autografts are only viable for patients whose skin can be safely taken from other areas of the body. Though effective, this treatment results in scarring and additional care is required for the areas from which the skin is taken for autografting. Another effective treatment is the use of artificial skin or skin substitutes but this is highly costly for patients who have large areas of burns.

In comparison to traditional methods, bioprinting offers a number of advantages including flexibility, reproducibility and shape retention. It can be applied to topical and transdermal formulation, studies of dermal toxicity and design of autologous grafts [3]. In the research by Lee *et al.*[3], a 3D skin tissue was constructed, which consisted of fibroblasts and keratinocytes for representing dermis and epidermis respectively, and collagen for representing the skin dermal matrix. After the 3D constructs were bioprinted, they were cultured in media conditions for maturation and stratification. It has been proved that the printed skin tissue was biologically and morphologically representative of real human skin tissue.

Currently, a stand-alone, rapid, autologous cell harvesting, processing and delivery technology entitled ReCell® Spray-On Skin™ from Avita Medical Ltd., Royston, UK, is designed for use in wound, reconstructive, burn and cosmetic procedures. This product enables clinicians and surgeons to treat skin defects using the own cells of the patient in a regenerative process. This accelerates the healing process as well as minimises scar formation whilst reintroducing pigmentation to the skin and eliminating tissue rejection. The ReCell® product has already been approved for marketing and sales in Europe, Australia and Canada, and is under clinical trials for marketing in the USA.

8.2.2 *Bone tissue engineering*

- Inkjet bioprinting

A study has demonstrated the feasibility of spatially controlling osteoblast differentiation and bone formation using the piezoelectric inkjet bioprinting technique [4]. The 3D patterns of bone morphogenetic protein-2 (BMP-2) were printed within human allograft constructs. The patterned growth factors are able to direct *in vitro* cell differentiation and *in vivo* bone tissue formation.

- Laser-assisted bioprinting: laser induced forward transfer (LIFT)

The LIFT bioprinting technique (see Fig. 3.28 and section 3.11.2) has been used in patterning of nano-hydroxyapatite and osteoblastic cells in two-dimensional (2D), and is now adapted to the 3D bioprinting of composite materials [5]. It was found that the physic-chemical properties, viability and proliferation of nano-hydroxyapatite were not altered and affected by the bioprinting process.

8.2.3 *Vascular graft*

Each year, thousands of "replacements" are needed: approximately 200,000 coronary artery bypass grafting operations and millions of procedures implanting interposition grafts to bypass obstructed arteries in peripheral vascular disease are performed in the United States alone [6].

Norotte *et al.* [7] reported a rapid and scalable extrusion-based bioprinting technique and applied it to fabricating tubular vascular grafts with a fine diameter. Various types of vascular cell such as fibroblasts and smooth muscle cells were aggregated into multicellular spheroids or cylinders, which were then printed along with agarose rods onto the substrate. Finally, the vascular tubes with a fine diameter were obtained as a result of fusion of the printed discrete units. This technique has shown a great potential to engineer multi-layer vessels with different shapes and complex hierarchical trees with branching geometries.

Skardal *et al.* [8] also employed an extrusion-based technique to directly print cellularised tubular tissue constructs made of hyaluronan hydrogels crosslinked with polyethylene glycol (PEG). The shapes of these constructs are very similar to simplified blood vessels. The bioprinted cell structures maintained a high viability in culture media for a month. This result offers hope that blood vessel structures and vascular networks may be bioprinted *in vitro* for clinical uses.

8.2.4 *Heart tissue engineering*

Duan *et al.* [9] synthesised a hybrid hydrogel made of methacrylated gelatin and methacrylated hyaluronic acid, which was bioprinted (extrusion) to form simplified heart valve conduits with leaflets and root. The hydrogels were first loaded into the dual syringes and extruded through the nozzle onto the horizontal platform. After the printing process was complete, the conduits were subjected to photocrosslink from the top to the bottom. The bioprinted heart valves were found to be able to maintain high viability. Thus, the bioprinting is a promising technology for tissue engineering of living valve replacements and investigating interactions of physiological valve cells.

8.2.5 *Cartilage tissue engineering*

There are around 1.2 million cartilage repair procedures conducted each year worldwide [10]. However, current cartilage TE methods are unable to engineer new tissue functionally similar to native cartilage in terms of extracellular matrix composition, zonal organisation and mechanical properties. Furthermore, in order to maintain long-term stability, the implants should be well integrated with surrounding and neighbouring native tissues. In the research by Cui *et al.* [11], human chondrocytes encapsulated in PEG-dimethacrylate (PEGDMA) were bioprinted through the use of simultaneous inkjet printing and photopolymerisation for repairing osteochondral defects. The compressive modulus of the bioprinted PEGDMA was found to be close to those of native human cartilage. Due to the simultaneous photopolymerisation, the bioprinted chondrocytes maintained the originally deposited positions, which ensured the precise cell distribution. High viability of chondrocytes was achieved and the implant steadily attached to the surrounding tissue. In addition, the bioprinted cartilage tissue also increased glycosaminoglycan (GAG) content. In addition to this, Roots Analysis Pte Ltd., released a report [1], in which 3D bioprinted knee cartilage replacements are expected to be available in the market within the coming decade.

Another group led by Fedorovich *et al.* [12] utilised fibre deposition technique to manufacture heterogeneous hydrogel constructs. Human chondrocytes together with osteogenic progenitors encapsulated in alginate hydrogels were bioprinted by using an extrusion-based bioprinter. It was revealed that the cell viability remained a high level throughout the entire bioprinting process. These findings indicate that it is feasible to fabricate viable tissues for repairing osteochondral defects.

8.2.6 *Bioprinted tissue models, drug discovery and toxicology*

The animal testing for research and drug testing has always been regarded as inaccurate and cruel. According to the data published by Human Society International [13], more than 115 million animals (mice, rats, guinea pigs, fish, birds, rabbits, farm animals, cats, dogs and non-human primates) are used and/or killed for experimentations in laboratory every year. Tissue constructs and disease models fabricated by bioprinting will open up an opportunity to test drugs on real human tissues as opposed to animals.

The first promising commercial application of 3D bioprinting technology is highly likely to be drug testing. A large amount of data has confirmed that many drugs fail during the late stages of development. Only 1 in 5,000 drugs finally reaches the market [14]. The overall process of drug discovery is lengthy, risky and costly. On average, it takes 12 years to develop a new drug and an investment of USD 1.2 billion [15]. Bioprinting can be utilised by pharmaceutical companies to identify major issues in drug development and thus lowering the risks and recognising potential drug failures at early stages.

Organovo Inc., San Diego, USA, is striving to launch a 3D liver model to be used in toxicity testing by December 2014, which will be the first bioprinted liver model available in the market. The liver model of 96-well format is under development [16]. Another company called Roslin Cellab, Edinburgh, UK, also aims to commercialise their liver model. Other micro-tissue models such as heart and kidney are also believed as potential opportunities in the market for drug testing.

Bioprinting enables heterogeneous tissues to self-assemble, mimicking physiological tumour model *in vivo*, and creating viable tumour models *in vitro* [17]. Pepper *et al.* [18] printed mammary tumour cells in predefined patterns. Moon *et al.* [19] encapsulated bladder muscle cells within highly viscous collagen and printed them using a extrusion-based technique. A co-culture cancer model containing ovarian cancer cells and fibroblasts was created by Xu *et al.* [20], however, no results regarding the functional characterisation of the model were obtained.

Researchers are actively developing various micro-tissues to support the application in drug screening, such as 3D cardiac micro-tissue models [21]. Matsusaki *et al.* [22] utilised the inkjet technique to print micro-arrays of tissue structures. Hundreds of micro-tissues were integrated into one single micro-array for mimicking actual cell-cell interactions. However, the biological properties of the cancer were not evaluated in this study. Dickinson *et al.* [23] fabricated breast cancer cell laden hyaluronic acid (HA) hydrogels, which were used to analyse the spatial mechanisms that regulated tumour angiogenesis.

References

[1]   R. Analysis, "3D Bioprinting Market: 2014 - 2030," 2014.

[2]   S. V. Murphy and A. Atala, "3D bioprinting of tissues and organs," *Nature biotechnology,* 2014.

[3]   V. Lee, G. Singh, J. P. Trasatti, C. Bjornsson, X. Xu, T. N. Tran, *et al.,* "Design and Fabrication of Human Skin by Three-Dimensional Bioprinting," *Tissue Engineering Part C: Methods,* vol. 20, pp. 473-484, 2013.

[4]   G. M. Cooper, E. D. Miller, G. E. DeCesare, A. Usas, E. L. Lensie, M. R. Bykowski, *et al.,* "Inkjet-based biopatterning of bone morphogenetic protein-2 to spatially control calvarial bone formation," *Tissue Engineering Part A,* vol. 16, pp. 1749-1759, 2010.

[5]   S. Catros, J. C. Fricain, B. Guillotin, B. Pippenger, R. Bareille, M. Remy, *et al.,* "Laser-assisted bioprinting for creating on-demand patterns of human osteoprogenitor cells and nano-hydroxyapatite," *Biofabrication,* vol. 3, p. 025001, 2011.

[6]   B. da Graca and G. Filardo, "Vascular bioprinting," *The American journal of cardiology,* vol. 107, pp. 141-142, 2011.

[7]   C. Norotte, F. S. Marga, L. E. Niklason, and G. Forgacs, "Scaffold-free vascular tissue engineering using bioprinting," *Biomaterials,* vol. 30, pp. 5910-5917, 2009.

[8]   A. Skardal, J. Zhang, and G. D. Prestwich, "Bioprinting vessel-like constructs using hyaluronan hydrogels crosslinked with tetrahedral polyethylene glycol tetracrylates," *Biomaterials,* vol. 31, pp. 6173-6181, 2010.

[9]   B. Duan, E. Kapetanovic, L. Hockaday, and J. Butcher, "Three-dimensional printed trileaflet valve conduits using biological hydrogels and human valve interstitial cells," *Acta biomaterialia,* vol. 10, pp. 1836-1846, 2014.

[10]   A. K. Leichman. (2013, 13/05/2014). *Good News for Knees.* Available: http://israel21c.org/health/good-news-for-knees/

[11]   X. Cui, K. Breitenkamp, M. Finn, M. Lotz, and D. D. D'Lima, "Direct human cartilage repair using three-dimensional

bioprinting technology," *Tissue Engineering Part A,* vol. 18, pp. 1304-1312, 2012.

[12]   N. E. Fedorovich, W. Schuurman, H. M. Wijnberg, H.-J. Prins, P. R. Van Weeren, J. Malda, *et al.,* "Biofabrication of osteochondral tissue equivalents by printing topologically defined, cell-laden hydrogel scaffolds," *Tissue Engineering Part C: Methods,* vol. 18, pp. 33-44, 2012.

[13]   HumaneSocietyInternational. (2012, 13/05/2014). *Animal Use Statistics.* Available: http://www.hsi.org/campaigns/ end_animal_testing/facts/statistics.html

[14]   On3DPrinting. (2013, 13/05/2014). *Bioprinting is a Multi-Billion Dollar Pharma Opportunity for 3D Printing.* Available: http://on3dprinting.com/2013/09/26/bioprinting-is-a-multi-billion-dollar-pharma-opportunity-for-3d-printing/

[15]   T. J. Philipson and A. v. Eschenbach. (2013, 13/05/2014). *FDA Reform Can Lift U.S. Economy.* Available: http://www.bloombergview.com/articles/2013-02-28/fda-reform-can-lift-u-s-economy

[16]   J. Napodano. (2012, 13/05/2014). *Zacks Small-Cap Research.* Available: http://scr.zacks.com/Theme/Zacks/files/August%2020,%202012 _ONVO_Initiating%20Coverage%20Of%20Organovo%20With %20Outperforming%20Rating_Napodano_v001_y49ps2.pdf

[17]   T. Boland, T. Xu, B. Damon, and X. Cui, "Application of inkjet printing to tissue engineering," *Biotechnology journal,* vol. 1, pp. 910-917, 2006.

[18]   M. Pepper, C. Parzel, T. Burg, T. Boland, K. Burg, and R. Groff, "Design and implementation of a two-dimensional inkjet bioprinter," in *Engineering in Medicine and Biology Society, 2009. EMBC 2009. Annual International Conference of the IEEE,* 2009, pp. 6001-6005.

[19]   S. Moon, S. K. Hasan, Y. S. Song, F. Xu, H. O. Keles, F. Manzur, *et al.,* "Layer by layer three-dimensional tissue epitaxy by cell-laden hydrogel droplets," *Tissue Engineering Part C: Methods,* vol. 16, pp. 157-166, 2009.

[20] F. Xu, J. Celli, I. Rizvi, S. Moon, T. Hasan, and U. Demirci, "A three-dimensional in vitro ovarian cancer coculture model using a high-throughput cell patterning platform," *Biotechnology journal,* vol. 6, pp. 204-212, 2011.

[21] C. Wang, Z. Tang, Y. Zhao, R. Yao, L. Li, and W. Sun, "Three-dimensional in vitro cancer models: a short review," *Biofabrication,* vol. 6, p. 022001, 2014.

[22] M. Matsusaki, K. Sakaue, K. Kadowaki, and M. Akashi, "Three-Dimensional Human Tissue Chips Fabricated by Rapid and Automatic Inkjet Cell Printing," *Advanced healthcare materials,* vol. 2, pp. 534-539, 2013.

[23] L. E. Dickinson, C. Lütgebaucks, D. M. Lewis, and S. Gerecht, "Patterning microscale extracellular matrices to study endothelial and cancer cell interactions in vitro," *Lab Chip,* vol. 12, pp. 4244-4248, 2012.

Problems

1. What are the major challenges for 3D bioprinting?
2. What are the four technical challenges for 3D bioprinting?
3. What are the two operational challenges for 3D bioprinting?
4. What are the two most promising 3D bioprinting techniques used in bone tissue engineering? What are the advantages of these two techniques over other techniques that make them ideal candidates for bone tissue engineering?
5. What are the application areas that 3D bioprinting can potentially be applied to in the future?

Index